T0250171

SCALE
HEAT | COOL

SCALE

HEAT | COOL

ENERGY CONCEPTS, PRINCIPLES, INSTALLATIONS

EDITORS
ALEXANDER REICHEL
KERSTIN SCHULTZ

AUTHORS
MANFRED HEGGER
JOOST HARTWIG
MICHAEL KELLER

Birkhäuser
Basel

EDITORS' FOREWORD

The individual books of the SCALE construction series have been chosen to represent different building design criteria as well as basic construction functions. This also applies to this volume - *Heat | Cool*. One of the basic human needs is that for protection against cold, and heat. In the history of buildings, these needs have led to the development of structures and systems ranging from the ancient open wood fire through to today's complex services installations. Within the possible range of temperatures, we perceive only the small range between 16 and 26 °C as being comfortable. Where this range cannot be achieved by natural means in our environment, we have to rely on artificial measures. The energy required for these measures is currently obtained from finite resources, whilst demand is increasing due to worldwide population growth. Therefore, architects and services engineers no longer need only to provide perfect ambient conditions in buildings as an additional element to the space design. Instead, both professions need to work together to produce an integrated design from the very beginning of the design process in order to develop intelligent energy-conserving concepts that contribute to minimizing energy consumption.

Energy efficient systems cannot be achieved using technical means alone. They need buildings capable of supporting such concepts in their functional layout, zoning, and structure. The design of a building, its materials, and building components affect the climate inside the building and hence its energy efficiency. The simplest way of avoiding excessive energy consumption is to abstain from using energy unnecessarily. This becomes apparent in self-sufficient buildings which provide a certain degree of comfort and coziness without technical means. The building masters of old, who did not have recourse to active systems, developed unique building patterns and aesthetics. For example, the shape of the roof of a traditional house in the Black Forest and the solid construction of its living quarters, in combination with the necessary hay loft and stables, contribute to a minimized heat requirement without the need for active energy. Similar principles are used in the Arabian atrium houses which achieve the necessary cooling through passive construction methods by lowering the buildings into the ground, providing an internal cooling water surface and a high tower for catching the wind.

For this reason, this volume of the SCALE series focuses not only on technical systems and installations, but uses selected completed projects to illustrate all the design stages involved, from the first sketches through to the last technical details. After familiarizing the reader with the relevant external and internal factors—from global weather data to use-specific internal heat sources and outputs—building structures and technical installations are covered in the respective chapters. The book deals with the technology involved in heating and cooling as well as mechanical ventilation and the generation of electricity. Special consideration is given to the design and creation of high quality energy concepts in order to emphasize the close interrelationship between the spatial design and its effect on the use of energy. One of our priorities was to explore avenues and initiatives for buildings that generate their own energy through a symbiosis of shape, structure, and system. Consequently, examples of current European and German regulations are included.

The case studies of recently built projects in the last chapter show that the energy component of buildings has now developed into a recognized quality that is characteristic in architecture. Appropriate use of resources and the balance between complex technology and simple building concepts are illustrated.
In this context, the architect develops his approach with regard to location, existing building methods and traditions, neighboring buildings, climate conditions, the building volume, openings, and the choice of materials. By integrating considerations of energy use into the building design at a formative stage, the authenticity of our cities and villages remains intact. At the same time, we are able to provide answers to questions about modern construction methods. In building refurbishment too, design parameters for optimization processes should be based on the appropriateness and understanding of the structures, substance, and qualities of an existing building. Likewise, building refurbishment for the purpose of improving energy efficiency includes an architectural dimension and should not be understood as just a technical add-on.

We hope that this volume will contribute to sustainably improving the quality of our built environment and lead to the development of a unique sensual architecture based on an approach that makes careful use of resources.

We thank the authors of *Heat | Cool* for their highly competent and committed co-operation in producing this volume, and the publisher, Birkhäuser—in particular Andrea Wiegelmann, for her dedication in producing this multi-volume series and her unfailing trusting cooperation.

Basel
1 August 2011
Alexander Reichel, Kerstin Schultz

HEAT | COOL
INTRODUCTION

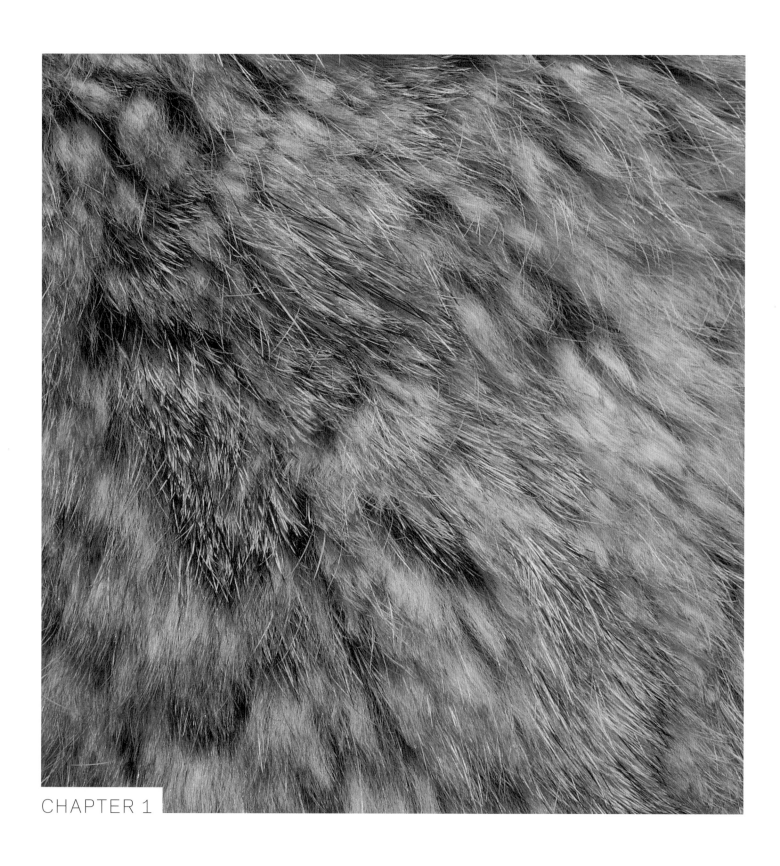

CHAPTER 1

ENERGY EFFICIENCY AND SUSTAINABILITY

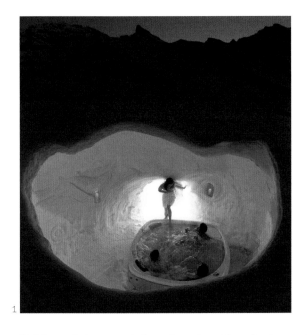

1

1   Whirlpool in the igloo village at Enegelberg-Titlis.

2   Office of the Institute of Building Services and Energy Design TU Braunschweig.

Buildings shelter people from the adversities of nature. We make our homes inside buildings as they provide a well-balanced environment for us. By protecting us from heat and cold, they preserve our personal health and increase our well-being.

The means to achieve this were developed and increasingly perfected over the long history of architecture. This process saw the emergence of construction methods, building types and technical devices very specifically matched to the local climatic conditions, cultural prerequisites and the available building materials. In our cool and moderate climate zones, solutions were required to provide shelter against the cold and wind, use solar irradiation to best effect and yet protect against too much radiation, while balancing out the extremes of seasonal differences in temperature. In contrast, buildings in dry/hot zones had to provide good ventilation and balance temperature differences, although in this case chiefly between day and night. In the tropics, the principal issues of architectural design are increased ventilation through the building as well as effective shelter from rain and sunshine.

These are just some examples of the general tasks a building must perform in different climate zones. Within these zones, there are further differences between construction methods. The geography and topography, conditions in the microclimate, the availability of building materials, and many other factors had an effect on building styles, which created houses for their respective locations that fulfilled their function as an "air conditioner" with the least possible input of resources.

The advent of the industrial revolution and the availability of cheap fossil fuels changed the way in which houses provide the required comfort levels. Comfort levels were significantly increased through improved construction methods, better building envelopes and, particularly, new technologies for heating, cooling, ventilation, and air conditioning of buildings; this contributed to better health and improved productivity of the occupants. However, these improvements came at a price as they needed more resources in terms of building work, materials, and technical installations. This increased use of resources has made the construction process largely independent of local conditions and has seemingly rendered the autochthonous building method superfluous, because now, any construction has become possible anywhere in the world, albeit subject to considerable expense in terms of construction and operation.

This expense is reflected in the energy balance of buildings. In temperate climate zones, about 40% of total energy is used for operating buildings, in particular, for heating and cooling. In Germany, heating alone is responsible for about 80% of the energy requirement of a building. This means that we are paying a high price for the success we have had in providing significantly increased well-being in our buildings. This relates to not only the high expenditure for technical installations and the operation of buildings, but also the greater dependence on resources from other, potentially unstable, countries and the justified concern that supplies are finite and that prices will keep rising.

Today, about 90% of the primary energy needed for operating our buildings is derived from fossil fuels. This means that the operation of our buildings and the resulting CO_2 emissions bear a large part of the responsibility for climate change, even more so than our transport system and industry, both of which are also large contributors.
Therefore, action is urgently needed so that we stop gambling with our future. Regarding the comfortable and healthy living conditions our society has achieved, it is not likely that there will be a readiness to forgo these achievements in the near future. But here is the good news: there is still significant scope for action and improving efficiency in construction.
In recent years, energy-efficient building has made considerable progress. Today, legal requirements are in place for new buildings, which means that these buildings only use one quarter of the energy used by buildings constructed before the first energy crisis. Whether these reductions are actually achieved in practice or not is yet to be ascertained. The means are available, however, to achieve this and much more. The Passivhaus standard, which is now applied fairly widely, requires that the total use of primary energy for the building is less than 120 kWh/m²a, and that for heating it is less than 15 kWh/m²a. This corresponds to just 7 to 8% of the energy used in buildings constructed during the eighties.

This is made possible by a considerable improvement in building technology, aided by appropriate thermal insulation, airtight construction, highly efficient windows and solar screening, storage mass, and highly efficient services installations. There are other factors however, that play an important role, such as efficient use of the available space while maintaining a high level of quality, a sensible building shape with appropriate zoning, sensible fenestration with a good balance between solid building fabric and transparent elements, and heat- and humidity-absorbing surfaces. But by itself, this vocabulary will not create a really energy-efficient house unless the elements are efficiently arranged to work together and complement each other. This book aims to cover both the vocabulary and the grammar.
With highly energy-efficient buildings it is also possible to generate any residual energy required for the building without damaging the environment, and to utilize the renewable energy available from the building's environment. Every building is surrounded by energy sources such as the wind and the sun, groundwater and soil, or biomass, the use of which saves resources and causes less pollution.

2

1

THE SOCIAL AND CULTURAL CONTEXT

The heating and cooling of rooms and buildings, while providing hygienic and health-promoting conditions, is intended to make it possible for people to spend long periods of time in them. In a working context it is also expected to contribute to an improvement in the performance of workers and the quality of products. However this should not just be seen as a purely technical concern, as this would not do justice to the whole issue. Users of buildings are more than sensitive sensors for heat and humidity; they need more than the supply of temperature, relative humidity etc. in certain acceptable ranges in order to satisfy requirements. Users exist in a social and cultural context which plays a significant role in determining the requirements as well as the solutions.

Ultimately it is not possible to objectively determine the tolerance range within which room conditions are perceived as comfortable. That this is determined quite differently in different countries is just one indicator of this fact. Let us take a look at the required air temperature in different social/cultural contexts using office workspaces as an example. They vary from the extremely narrow tolerance range of just +/- 10 °C in North America via a range of up to +/- 40 °C in Europe to ranges in Asian countries, including those in tropical climate zones, which are not regulated at all. To explain this just on the basis of differences in technological development would not be doing justice to the issue.

PERCEPTION AND PHYSIOLOGY

Let's dwell on room air temperature and consider the behavior of people in conditions that are often way beyond the tolerances stipulated by laws and standards. Even where temperatures are near freezing point, we can see that people work and stand around chatting in railway station halls, tents, conservatories, or temporary buildings. These people compensate for extreme temperatures with their clothing. Or they utilize the heat of the sun's radiation, which heats their bodies irrespective of the air temperature. We can observe similar behavior at the opposite end of the scale: even in extreme heat, people work in parks, on balconies, and in well-ventilated rooms. They make use of the fact that the passing wind cools their skin through the resulting evaporative cooling, which can create a more comfortable sensation than that perceived in many rooms with significantly lower room air temperatures.

In other words: on the one hand, requirements for relative room air humidity, surface, or room air temperature can provide important design parameters under standard conditions. On the other hand, such parameters do not fully represent the complex relationships, such as that between room air temperature and radiation, or human needs and preferences that change in different situations, such as in different weather conditions. Likewise, physiological requirements cannot be pinned down to objective standard specifications. People who spend a lot of time in standard room conditions that do not provide much stimulation may seek to break out of this environment, to feel the sense of coldness, to breath more deeply, to expose themselves to the first rays of sun in spring or the caressing wind on a hot summer's day. These too are justified user needs. It is one of the special tasks of architecture to provide answers beyond the satisfaction of objective requirements for heating and cooling.

CITY AND SPACE

Just as people from certain cultural and social contexts have evolved specific ways of life, so they have established their buildings and cities over long periods of time. Very often they have responded to the climatic conditions described above and gradually developed city spaces, building types, and construction methods that reflect their local conditions. They also reflect the specific social and cultural conditions and the availability of resources and materials. Today, this and many other factors affect every room, every city, and its buildings in their own unique way.

Design and construction, which seek to satisfy demanding requirements for heating, cooling, and energy efficiency, do not happen outside the above frame of reference. Rather, they take the thread of architectural

1 Interior of Gründerzentrum Hamm-Heessen, 1998, HHS Planer und Architekten.

2 Ningbo historic museum, 2008, Amateur Architecture Studio.

development and continue spinning it in a creative way. In this context, we may find important and helpful pointers in the architectural heritage of vernacular construction from the time prior to the availability of cheap fossil fuels and materials in a global market.

A new building culture of sustainable building will take its cue from its immediate environment, but without falling into the trap of romantic cliché or eclecticism. In its unbiased search for new solutions it will develop new, strong, and fascinating expressions of sustainable building that is fit for the future. And fit for the future it should be—ideally already anticipating future requirements today.

TASKS AND SOLUTIONS

In new buildings, sensible principles of such an approach to building can be creatively implemented and can contribute to creating room climates, while complying with energy requirements without the need for excessive technical services input. Here especially, the architectural task should take precedence over the technological solution: as a rule, the service provided by a building is cheaper and more long-term than that of its technical installations. The essential elements of architecture are form, mass and transparency, materials, textures and colors, as well as functional and typological considerations. Used properly these elements are extremely effective for resolving the issues surrounding room climate, energy economy, and climate protection.

Today, the improvement of existing buildings already constitutes the major part of all construction activities. In this context it is important to weigh up whether a building is a valuable part of the cultural heritage and, in the eyes of the population as well as experts, is instrumental in imbuing a city or neighborhood with a certain character. At least in the public realm, the appearance of such buildings should not be changed by energy-efficiency upgrade measures. In this case it would be advisable to apply a gentle approach to the efficiency upgrade of the building fabric while at the same time optimizing the building technology. Where this results in deficiencies in terms of room climate, this should be tolerated. Where such a building's contribution to climate protection is reduced, it may be possible to provide alternative energy saving measures in the immediate environment as compensation.

Buildings and building groups designed in everyday vernacular, the appearance of which does not reach the threshold of positive perception of a city's population, can be refurbished and upgraded without preserving their external appearance. Like new buildings they can make an effective contribution to energy conservation and climate protection while providing better comfort levels to the residents and addressing running costs as well as worries about the future.

Nevertheless, one basic aim for the urban built environment should be upheld: no change without beautification. Building is not something that just fulfills objectives which are defined one-dimensionally. We will not succeed in finding a really sustainable solution to the challenge of energy-efficient building unless we take the whole picture into account, including respect for our architectural heritage.

2

BASIC PARAMETERS

The need for heating and cooling buildings arises from the discrepancy between the climate conditions prevailing at a location and people's physiological requirements. In an ideal case, the user will perceive the predominant climate as comfortable or he will adapt his understanding of comfort to the external conditions. In these rare cases, there is no need to heat or cool buildings. In most cases however, climatic differences between indoors and outdoors have to be coped with, at least on a temporary basis. In areas with distinct seasons, external conditions change during the course of a year. In Central Europe the average external temperature can drop below -10 °C in winter while in summer, temperatures of over 40 °C are common. Other regions, such as deserts, are subject to similarly extreme fluctuations from day to night. In contrast, the requirement for comfort remains largely the same. Depending on their activity and the relative humidity, the temperature range considered agreeable by people is between 16 and 26 °C. It is possible to reduce the difference to the ambient temperature through appropriate building measures, for example, thermal insulation to prevent cooling, solar screening to prevent overheating, and storage mass to balance out the peaks of temperature. If these passive measures forming part of the building fabric are not sufficient, the temperature can only be maintained at the required comfort levels by using technical services components. Nevertheless the aim should be to avoid, as far as possible, the active input of energy and where required, design services installations to be as efficient as possible. This calls for an energy concept to be developed and taken into account in the building design at an early stage. Only when both of these design concepts work together will it be possible to arrive at an appropriate and specific architectural form.

and quantity of precipitation affects relative humidity. This and the predominant wind directions ⤳ **p. 28** affect the way in which the structure of buildings is protected against moisture. In addition, there are more local factors affecting the climate, such as buildings in the neighborhood (overshadowing and reduced solar irradiation), water surfaces and green spaces (cooling through evaporation). ⤳ **p. 30** These urban structures are the result of historic development as well as of legislation and planning regulations. ⤳ **p. 36** Likewise, these structures also determine the available utility services such as gas or district heating supplies ⤳ **p. 34**, and hence the decision for or against certain fuels. Last but not least, regional architectural features and local artisan traditions that impact on the appearance of a city or landscape can provide us with clues for finding forms of building construction suitable for the climate and careful with the use of resources.

EXTERNAL FACTORS
In order to develop design-integrated energy concepts, the designer must be familiar with the external factors prevailing where the building is to be erected. One of these is the climate. The angle of solar radiation depends on distance from the equator; this affects how much and which sides of the building envelope are reached by the rays of the sun. This determines whether a building is suitable for generating solar energy (e.g. photovoltaics) and whether there is a need to provide screening devices to openings in order to avoid overheating. Daily sunshine hours and radiation intensity vary more from summer to winter in locations further away from the equator. ⤳**1** The amount of solar irradiation is a deciding factor in determining the external temperature throughout the seasons and throughout the day ⤳**2**, and thus also the need for building insulation in order to maintain the internal temperature at a comfortable level. The frequency

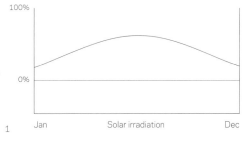

1

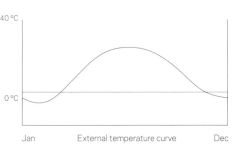

2

1 Typical annual course of solar radiation in Central Europe. Due to the distance to the equator and the angle of solar irradiation, the amount of radiation is never the same as that at the equator (100%). Due to the tilt of the earth's axis and the resulting seasons, the angle of irradiation changes throughout the course of a year and the radiation intensity fluctuates accordingly. During the cold season in particular there is less solar irradiation.

2 Typical annual temperature curve for Central Europe. The average temperature fluctuates throughout the course of a year due to changes in the intensity of solar irradiation. While external temperatures in summer are usually comfortable, in winter buildings have to make up the difference between internal and external temperature, and maintain the balance.

Energy balance of buildings:
3   Parameters of an energy balance:
a   Barrier to the outside
(building envelope)
b   Barrier on the inside
(wall between two different
temperature zones)
c   Heat flowing to the outside (ventilation and transmission heat losses)
d   Heat flow between different areas
e   Internal user heat sources
f   External input energy (solar irradiation, heat from heat-generating
appliances, etc.)

4   Where there is an imbalance between two rooms or between the
inside and outside, energy will flow
from the higher energy level to the
lower (from the warmer room to
the colder room, or outside). Without additional heat input the temperature difference between rooms
will gradually disappear. The time
it takes for the different temperature levels to balance out is determined by the thermal resistance of
the walls and by air changes.

5   Ideal scenario: The external factors are such that the conditions inside the building fulfill comfort requirements. There is neither heat
gain (sources) nor heat loss (heat
sink). This condition only prevails
for longer periods in a few climate
zones, like the savanna. In other climate zones this condition prevails
temporarily during the course of a
day or year (transition periods).

6   The energy in the building will
balance out of its own accord, depending on the external and internal heat sources and heat sinks, the
quality of the building envelope and
the ventilation. In this process it is
possible that rooms with strong internal heat sources will overheat
and others will cool down. The imbalance between rooms and the
outside can be minimized with
building measures, which are also
called passive measures. Synergies
can be created by linking up rooms.

7   The desired temperature level
can only be maintained with active
and controlled input and output
of heat. For this purpose it is
necessary to continuously measure
the current condition and compare
it with the target condition. Each
active interaction will result in an
energy requirement. The question
is whether additional heat needs to
be generated, or whether heat just
needs to be transferred from one
room to another. It is also possible
to link up spatially separated zones
by technical means.

3

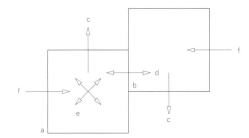

4

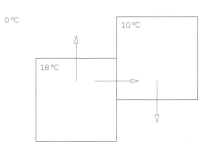

5

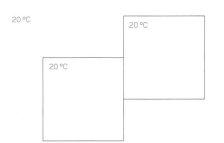

6

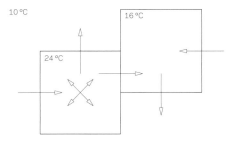

7

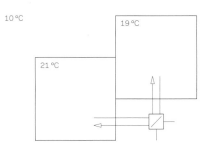

INTERNAL FACTORS

The type and intensity of use of a building result in a number of internal factors that also have to be taken into account in the design. People require certain comfort levels
→ p. 46, while machinery and production processes also
necessitate certain room climates to comply with technical requirements. → p. 50 The provision of heat uses up
energy, which, after it has been used as an internal heat
source, affects the thermal balance of the room. In addition, other resources such as oxygen are used, which then
has to be emitted as CO_2. Information such as operating
hours, occupation density, temperature and humidity requirements is compiled in so-called use profiles. → p. 52
For this purpose it is necessary to carry out an analysis
of the use and organization of the building in terms of
space and time. → p. 62 Based on these use profiles it
is possible to sub-divide the building into zones. For example, related rooms with an identical function can be
allocated to a zone provided the external factors permit
this and do not require a division due to their different
orientations within the building.

Once the zones have been determined and external and
internal factors have been established, the energy balance of the building can be analyzed. In order to arrive at a
thermal balance, it is necessary to study at least two conditions (outside and inside) and the type and quality of the
building envelope separating these conditions. There are
thermal and material flows between the inside and outside, such as air changes, heat loss through building components, and solar heat gain through windows. If there is
an imbalance between the temperature inside and outside, heat flows from the warmer side to the colder side.
If there is no movement of heat, the system approaches
a balanced state. If a building remains unheated, it cools
down to ambient temperature over time.

Once the external and internal factors have been established it is possible to identify existing deficiencies and
potentials. Any such deficiencies should be optimized
by improving the organization and design of the building
and conversely, by taking advantage of unused potential
in the same way.

BUILDING

How quickly a building adapts to the external temperature depends on many factors, most of which can be influenced by the designer. A compact building shape minimizes the total outside surface through which heat can
be transmitted → p. 68, reducing the amount of heat lost
compared to a building with a larger outside surface. In
addition, it is possible to slow down the loss of heat by
improving thermal insulation. This may result in extra
building components and impact on the appearance of
the building. → p. 77

Windows can be used to absorb energy into the building from solar radiation (solar gains). ↘ p. 82 This makes it possible to counteract the loss of energy through the building envelope in winter, at least during sunshine hours. On the other hand, solar gains during the summer months can lead to excessive heat in the building, which is perceived as uncomfortable. Organizing the layout of a building to take account of the movement of the sun makes it possible to use solar radiation for heating the interior (on the side facing the sun) or conversely, to avoid additional heating (on the side facing away from the sun). ↘ p. 70 The amount of solar irradiation can be influenced by the size and arrangement of openings as well as by solar screening measures. ↘ p. 82 Different types of solar screening devices are available to suit different orientations towards the sun. Measures relating to building components and building design are frequently referred to as *passive* whereas technical installations are called *active* measures.

Through the use of synergies within the building, it may be possible to use excess energy from one zone to cover the energy requirement of another zone. An example for taking advantage of unused potential is the body heat given off by pupils in a densely occupied classroom. If the facade of such a room is highly insulated, this excess heat may be used to maintain comfortable temperature levels in the building without additional heat generation until well into the cold season. The linking up of different zones of a building increases comfort and reduces energy requirements: in this way a zone can be prevented from overheating and the excess heat used for heating other zones where heat is required. The linking of temperature zones may even become an organizational principle for the building and be adopted in the architectural design.

Using heat storage in the form of storage mass ↘ p. 74 makes it possible to store heat built up during the day for utilization after a time delay, e.g. next morning. This consideration affects the designer's choice of whether a building has lightweight or solid construction.

TECHNICAL INSTALLATIONS
Once all passive measures and synergies have been exploited, there is usually a residual need for energy to make up the thermal balance. This means that external services are needed to run appliances such as boilers or cooling aggregates. These are needed to provide heating, cooling, and/or ventilation in order to achieve the required comfort levels. These systems can be integrated into building components (surface heating and cooling, building component activation, ventilation outlets) or they can be installed as distinct equipment (radiators, ceiling

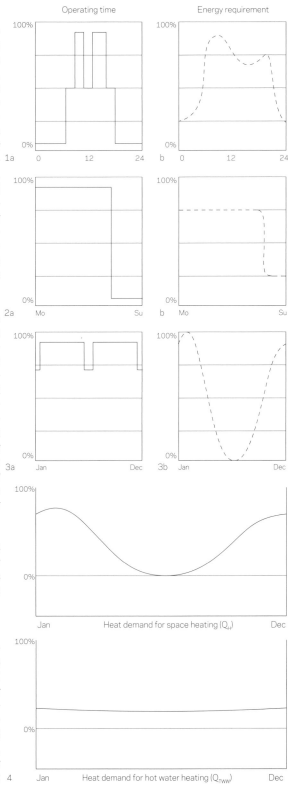

Operating time — Energy requirement

1a   Operating time: 100% 0% | 0 12 24
b   Energy requirement: 100% 0% | 0 12 24

2a   100% 0% | Mo — Su
b   100% 0% | Mo — Su

3a   100% 0% | Jan — Dec
3b   100% 0% | Jan — Dec

100% 0% | Jan — Heat demand for space heating (Q_H) — Dec

100% 0% | Jan — Heat demand for hot water heating (Q_TWW) — Dec

Operating times analysis:
1 The heat balance and resulting energy demand of zones fluctuate, particularly in non-residential buildings. Offices are typically used from 7:00 a.m. to 7:00 p.m. during the course of a day and remain empty during the remainder of the day/night. There is a constant change between times of overheating and cooling; peaks of high and low energy demand alternate.

2 There are also changes during the course of a week: during working days, non-residential buildings are nearly fully occupied whereas at weekends they are almost completely empty. When there are only a few persons using the building, it may be possible to heat/cool only certain rooms and in that way avoid having to heat/cool the whole building. A high standard of building construction minimizes the effects of cooling and shortens the time it takes to heat up the building.

3 Over the course of a year there are holiday periods as well as public holidays when the occupation density is reduced. The reduction of internal heat sources leads to higher heat requirements in winter but also reduces the risk of overheating in summer.

4 Annual energy requirement: Energy requirement results from the balancing of external and internal heat sources and sinks (losses and gains) and determining the remaining amount of energy needed by a building in order to maintain the required interior comfort levels. Common parameters are:
- annual heat requirement Q_H
- heat requirement for domestic hot water Q_{TWW}
- annual cooling requirement Q_C
- annual primary energy requirement Q_P
The energy requirement of a building can vary considerably depending on the factors described. Heating and cooling, as well as solar irradiation, are subject to big seasonal changes while domestic hot water and electricity consumption are steadier. The annual value of the energy requirement of a building does not provide enough information to design an energy concept or to determine the energy balance of a building at a certain point in time. For this purpose, a more detailed study over time is required as well as a graphic evaluation.

Energy requirement and power: Dividing the energy requirement into monthly values shows the approximate seasonal behavior of the building; daily and hourly values provide further detail. If one connects the individual values with each other, a curve results. The curve indicates the energy input required over a period of time in order to ensure comfort levels. In physics, energy divided by time is called power; in the construction industry the terms *heat load* and *cooling load* are commonly used. The area below the curve can be called the energy requirement, because power multiplied by time equals energy. The following applies: the shorter the observation period, the more precise is the understanding of the energy balance in a room or building. The heat load and cooling load can be used to evaluate the quality of a building, and also as input parameters for the design of appliances.

5   The total heat requirement is computed in annual curves; here it includes the sum of the heat required for space heating and domestic hot water.

6   The yield from a solar thermal installation is also subject to seasonal changes, albeit counter to the demand for heat. When the external temperatures are very low it is not technically possible to convert solar irradiation into a useful yield.

7   Superimposition of total heat requirement and available solar heat. The diagram shows the proportion of energy requirement which can be covered by solar energy as well as the energy deficit in winter and the unused heat excess at the height of summer. The relationship of these areas to each other provides an indication of other parameters, such as solar coverage (yield to demand).

8   During summer, any heat generated in excess of demand dissipates unused unless there is a seasonal storage facility (e.g. a water tank) so that the supply can be used to meet the demand at a later date.

9   A second controllable heat generating device covers the remaining energy requirement. Since it is not possible for a solar thermal collector to harness solar yield during winter, the heat generating device has to be designed for the maximum required heating output (heating load on the coldest winter day). Such a unit is called a peak load generating device. This calls for technologies with a low proportion of investment per kW output.

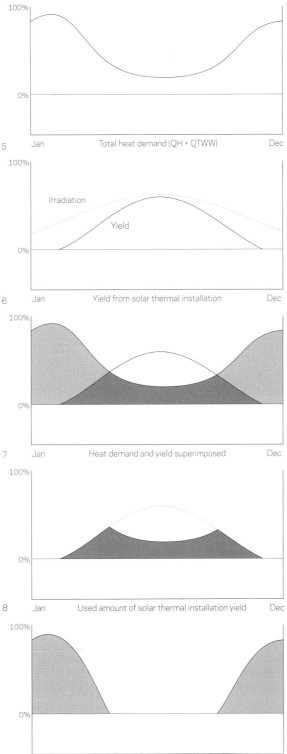

5   Jan   Total heat demand (QH + QTWW)   Dec

6   Jan   Yield from solar thermal installation   Dec

7   Jan   Heat demand and yield superimposed   Dec

8   Jan   Used amount of solar thermal installation yield   Dec

9   Jan   Peak load of heat generating device   Dec

radiators). ➘ p. 96 Water- or airbased distribution lines are used to connect the transfer systems to the heating and cooling appliances. ➘ p. 102 Buffer storage decouples the time of heat generation from the time of its use. ➘ p. 104 Not having to respond to peak demands quickly in real time, the generating units can operate at a lower output for longer periods, which means that they can be smaller. There are many different generating systems to choose from. ➘ p. 106 They vary in terms of design, the fuel used, the level of temperature provided, as well as the cost of investment related to performance. The choice of system determines the type of primary energy used. ➘ p. 116

It is important that all technical installations are integrated into an overall concept not only by using energy synergies between zones already described above, but also by taking advantage of technical synergies. Taking a school as an example: classrooms need a high input of fresh air in order to ensure adequate air quality. If a controlled ventilation system is used, then heat can be recovered from the exhaust air and used to heat the fresh incoming air. In this case no additional ventilation equipment is needed but it is simply a case of making best use of the existing equipment for servicing the room. Furthermore it is possible to prevent overheating in summer when internal and external heat sources and heat sinks do not balance, but work cumulatively with the resultant negative effect on the thermal balance of the room.

ENERGY CONCEPTS

Energy concepts take account of internal and external factors, producing an evaluation and design recommendations for the design of the building envelope and technical services. ➘ p. 18 The objective is to avoid the need for heating and cooling where possible, to take advantage of unused heat sources and heat sinks, to minimize the remaining energy requirement, and to cover this requirement from renewable energy sources.

DEVELOPING ENERGY CONCEPTS

Energy concepts can be developed for new buildings as well as for buildings awaiting renovation. The following example shows how the above-mentioned aspects (internal and external factors, the building, and technical installations) are combined with each other in a comprehensive energy concept for an existing building.

The example used is a high school at the edge of a town in southern Hessia. The building was constructed in 1972 and, six years later, a second building volume was added. There are 70 teachers teaching approx. 650 pupils.

The study starts with a thorough analysis of the prevailing conditions and factors, deficiencies, and potentials, which are then utilized, remedied, or enforced as the design proceeds.

EXTERNAL FACTORS

The climate in southern Hessia has the following average values: the average annual temperature is approx. 9 °C. The average temperature in January is around 0 °C (lowest value), and in July, approx. 20 °C (maximum value). ⟶1 In summer the temperature can vary between 20 °C and just under 40 °C during the course of a day. On cold winter days, the temperature fluctuation is between -20 and 0 °C. The average precipitation is 60 to 65 cm/ m². ⟶2 The region receives global radiation of 1,000 to 1,050 kWh per square meter per year. Again, there is a significant difference between summer and winter. The school building is surrounded by dense vegetation, which minimizes or prevents solar gains.

The school is beautifully situated in the midst of a park, which could be considered as a potential. It has a range of options for providing attractive areas for break periods, possibly by agreement with the other surrounding schools. As there are other schools in the neighborhood, one option would be to consider a refectory for the pupils.

INTERNAL FACTORS

The school has a number of deficiencies. ⟶3 Owing to serious defects in the facade and roof, the building has suffered moisture damage; in addition, much heat is lost through the building envelope (transmission heat loss). Both windows and opaque building components are far below current insulation standards. Interior comfort levels are badly affected by poor air and light quality, and the sound insulation between classrooms is inadequate. The obsolete heating and ventilation system is faulty and does not work properly, and this results in high operating costs. Nobody on site is familiar with the technical installations, which makes maintenance and servicing difficult. Large internal areas have to be lit with artificial lighting, which results in enormous electricity costs and an unpleasant ambiance. The lack of classrooms and break/staff rooms affects the quality of the accommodation. Pupils need spaces where they can take breaks and spend their free periods. They are not prepared to take ownership of areas and look after them unless the school is attractive for them in terms of living space.

Both building tracts are compact in design, which results in a good proportion between surface and volume. The primary load-bearing structure of the building is intact and can be retained.

1   Monthly average temperature [°C].

2   Monthly average precipitation [cm/m²].

3   Analysis pictograms:
a   leaky building envelope
b   extremely high electricity consumption
c   obsolete/faulty equipment
d   lack of identification
e   poor sound insulation

4   Pictograms showing solutions:
a   atrium
b   modified layout
c   more daylight
d   new facade construction
e   circulation areas more attractive

5   First floor layout before, not to scale.

6   First floor layout after, not to scale.

7   Calculation of areas.

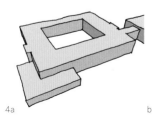

3a

b

c

d

e

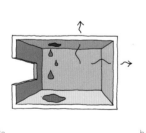
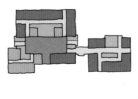

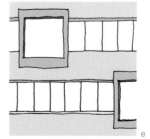

4a

b

c

d

e

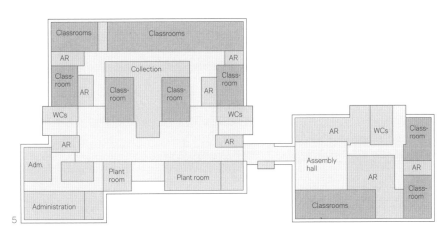

5

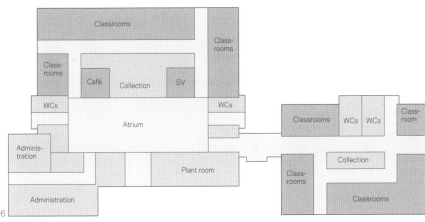

6

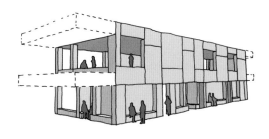

As a rule, the school building is used from early morning to the afternoon. At weekends and during the holidays, the building does not require heating or cooling. For periods like these, it would be advisable to consider sharing the use of the building with other institutions, such as adult education or music schools.

The above factors are such that everyday teaching is affected. The objective of the renovation is to provide pupils with a comfortable and stimulating learning environment in which they can study with good concentration and relax during the breaks. The following measures were identified. ⤿ 4

BUILDING MEASURES

A central element of the design and refurbishment concept is the newly created two-story roofed atrium. It is inserted at the center of the western part of the building. ⤿ 5, 6 It creates an open area of about 1,000 m² on the ground floor. A cube-shaped object has been placed into the atrium in order to break up the space. Similarly, the center of the eastern part of the building features such an object. It houses collections and the library. The central enclosed area houses the repository and the outer areas the reading section. The centrally positioned café for pupils enlivens the center of the complex, and the office of the pupil representatives (SV) is also centrally positioned so that it is readily accessible to all pupils. Large openings in the facade of the core building open up many vistas to the library, SV office, and pupil café. An exception is the side facing the assembly hall. It is uniform and largely without openings as it provides a back wall with a stage for the assembly hall. The timber surfaces of the two new core buildings distinguish these from the other parts of the building complex with its different materials. A new saw-tooth roof with transparent slopes facing north and opaque slopes facing south creates even lighting and protects against overheating and glare.

The narrow sides of the atrium are enclosed on the upper floor and feature fire protection elements which, in the case of fire, divide the building into two sections.

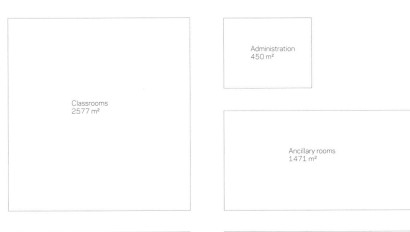

Classrooms
2577 m²

Administration
450 m²

Ancillary rooms
1471 m²

Break hall/atrium
1014 m²

Access
1128 m²

7

1

FUNCTIONAL ARRANGEMENTS

As part of the building redesign, the internal functions were also rearranged. The entrance area, now two stories high, leads to the central assembly hall. On normal teaching days the assembly hall is used by pupils during their break time, but it can also be used for special events and can be equipped with a stage. The balcony on the upper floor opens up vistas to the central hall and the other buildings. The classrooms are arranged along the facade, where they benefit from good natural lighting and only ancillary rooms with temporary use are located internally. The first floor also accommodates the natural sciences teaching rooms, the use of which is more flexible. The collections are housed in internal rooms. The single-story extension to the southwest provides accommodation for administration. On the second floor of the core building is the library, with classrooms arranged around it. The eastern part of the building comprises classrooms along the facade and also a core with ancillary rooms. The stairwells, toilets, and services rooms are retained.

IMPROVEMENTS IN THE DESIGN

The design of the central cores is aimed at improving the quality of the access areas. While the facades facing the corridors are rearranged with a number of recesses with benches to form seating niches for use by pupils during their breaks and free periods, the corridors extend through to the facade, where they are emphasized on the outside by large wooden structures. This provides natural lighting to the corridors and also creates vistas to the park. The timber frames are also used for benches in order to create areas of open communication in the building. Recessed fenestration bands provide an opportunity for fitting external solar screening louvers at the intermediate level on the south, east, and west of the building. Where the glazing is located near the outside face of the elevation because of the seating recesses, solar protection glass is used.

The new visible facade comprises vertical timber siding and is moved further outwards in order to gain space and avoid thermal bridging at the faces of the concrete slabs.

The timber conveys a warm and natural feeling, and the horizontal fenestration bands are interrupted by the timber structures with their colorful paint finish.

All windows are replaced. In order to improve daylight levels in the building, the vegetation around the building is reduced. The use of artificial lighting is further reduced by light paint and floor finishes.

TECHNICAL INSTALLATION MEASURES

In addition to the above building measures, the upgrading work also focused on the technical installations.

The saw-tooth roof above the atrium is designed in its orientation and slope so that it can be used for the installation of a photovoltaic system. ⟍1 In the western part of the building there are twelve south-facing roof areas sloping at a 30° angle, all of which can be fitted with photovoltaic elements. On the western part of the building a total roof area of 359.42 m² is available and on the eastern roof there is another 104.83 m², making a total of 464.25 m² of usable roof area.

The school's existing warm air heating and ventilation system is retained for further use. ⟍2 Likewise, the connection to the district heating system, which is operated by a combined heat and power plant (biogas), is retained and is used to operate the modernized heating system. In combination with the heat gained from the heat recovery in the ventilation system, it is possible to substantially reduce the heat requirement. In addition, the intake air is conducted through an earth duct, which means that it is pre-warmed in winter and pre-cooled in summer. The external air intake aperture is situated to the north of the building, where it is overshadowed by the building. Running along the balcony of the main building, the ventilation ducts deliver the air to the atrium and classrooms from a mixed air ventilation system and also take out the exhaust air. As the facade is now well insulated, the heat emitted from the pupils can be put to good effect so that, as a rule, the heating system can be switched off by mid-morning once the building is up to temperature.

The solid floors provide sufficient storage capacity to be able to absorb excessive heat during peaks in summer.

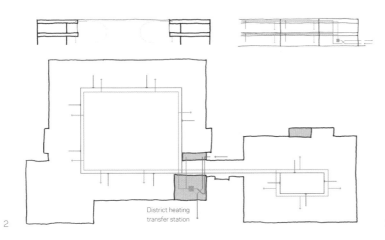

District heating transfer station

2

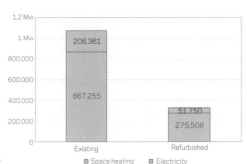

3

2 Ventilation schemes.

3 Perspective of atrium.

4 Comparison of primary energy consumption (in kWh/a).

5 Savings potential of the different measures (in kWh/a).

6 Comparison per person – before/after.

This balancing potential is retained during heat waves by allowing the building fabric to be cooled down overnight with nighttime ventilation.

Both the building measures and technical installation measures are supported by changes in organizational arrangements. Previously, all break periods were ten minutes long, which meant that pupils were not able to make use of the building and its environment during their breaks. By rearranging the duration of break periods, pupils enjoy longer breaks and have the opportunity to make better use of the surroundings. The change in the length of teaching and break periods made it possible for the pupils to take advantage of the improved break areas.

In addition, it was possible to reduce energy consumption by arranging the layout of the various rooms more appropriately.

Overall, the refurbishment and upgrade of the building has resulted in significant savings. ➔ 4-6 The energy requirement for heating and electricity has been reduced to about one third of the original.

At the same time the architectural quality of the building was significantly enhanced, as can be seen in the diagram and modified layout plans. There is now more space, more air volume, and more window area per person.

This example illustrates that an energy concept needs to combine aspects of space and function, of building construction, of technical installations, and of social interaction in order to achieve a holistic architecture. In this way the energy concept combines with the architectural concept to facilitate buildings that are fit for the future.

4

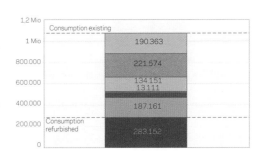

5

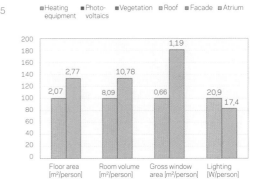

HEAT I COOL
EXTERNAL FACTORS

CLIMATE / MACROCLIMATE

The climate of a location is clearly one of the external factors affecting the design of buildings and, more particularly, the energy concept employed.

The term *climate* refers to possible weather conditions that occur in a location over a long period of time, including typical sequences of daily and seasonal fluctuations. The climate of a location is affected by climate factors and climate elements.
Climate factors include solar radiation, the ratio of land to sea (continental climate/maritime climate), the altitude above sea level, but also consequences of human activity (forest clearance, emissions) that contribute to the creation and variability of the climate or help to maintain it. Climate elements are meteorological parameters that interact with each other and characterize the climate. Climate elements include air temperature, humidity, wind, precipitation, degree of cloud cover and duration of sunshine, as well as their interaction. Climate elements are subject to daily and seasonal fluctuations. Climate factors and climate elements define a climate system, which

in turn is affected by external factors (topography, tilt of the earth's axis, rotation, and revolution of the earth, volcanism).
Distinctions can be made between macroclimate, mesoclimate, and microclimate, depending on the level of study.

MACROCLIMATE
A climate always relates to a particular location although, on a global level, there are large regions with similar climate conditions. These climate zones make up the macroclimate of the earth. A distinction is made between seven climate zones: tropical rainforest, savanna, steppe, desert, temperate, cold temperate, and tundra climates. ↘2
These different climate conditions give rise to different requirements for buildings. ↘p. 26 Traditional regional building methods demonstrate architectural options for meeting these requirements using locally available resources and technologies, and as a rule achieve high comfort levels as well as distinct styles through relatively simple means. ↘1

1 Traditional methods of construction in different climate zones. Construction methods depend on climate conditions as well as locally available materials, construction methods, and the lifestyle of the population (settled or nomadic).
a Rice stores in Indonesia (rainforest climate)
b Chinese roundhouse (tropical savanna climate)
c Arabic wind towers (desert climate)
d Mongolian yurt (steppe climate)
e European timber frame house (temperate climate)
f Thatched cottage in Ireland (temperate climate)
g European stone house in the high mountains (temperate climate)
h Igloo in the Arctic (tundra climate)

1a

b

c

d

e

f

g

h

2   The earth's climate zones and typical annual temperature and precipitation curves. The temperate and cold temperate climates can be further subdivided into *dry* and *humid*.

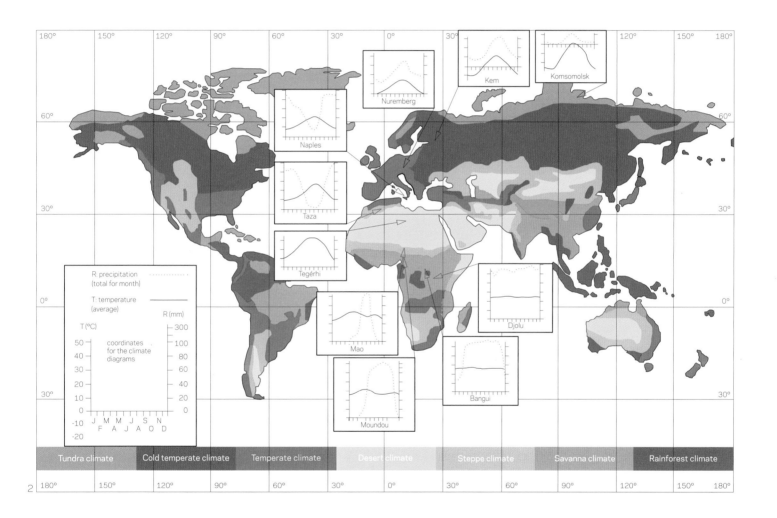

Description of climate zones, their predominant climate factors, and elements and resulting construction requirements for buildings in these climate zones.

Climate zone	Climate conditions	Building requirements
Tropical rainforest climate	No distinct seasons, heavy precipitation, monthly average temperatures 25 to 27 °C, daily temperature fluctuations more pronounced than seasonal fluctuations, high humidity levels (hot and humid)	Protection against precipitation, enhancing natural ventilation
Savanna climate	Rainy season alternates with dry season, high levels of precipitation during the rainy season, monthly average temperatures 23 to 27 °C, daily temperature fluctuations more pronounced than seasonal fluctuations	Protection against precipitation (rainy season), protection against solar irradiation (dry season), enhancing natural ventilation
Steppe climate	Short rainy season with heavy precipitation, which sometimes fails to occur, primarily dry, more evaporation than precipitation, monthly average temperatures 0 to >30 °C, high daily and seasonal temperature fluctuations, low humidity, dust storms and blizzards	Protection against precipitation (rainy season), protection against solar irradiation, making use of storage mass, enhancing natural ventilation at night (nighttime air cooling), protection against strong winds
Desert climate	Dry, sometimes no precipitation for years, monthly average temperatures between 15 and >30 °C, high daily fluctuations in temperature of over 50 K, dry air	Protection against solar irradiation, making use of storage mass, enhancing natural ventilation at night (nighttime air cooling)
Temperate climate	Dry summers and humid winters (in the south, e.g. Mediterranean area), always humid (in the north, e.g. Germany), monthly average temperatures from <0 to 25 °C, daily temperature fluctuations less than seasonal fluctuations	Protection against solar irradiation (summer), making use of solar gains and protection against heat loss (winter and transitional seasons), making use of storage mass
Cold temperate climate	Always humid or, more to the north, winter dryness, monthly average temperatures of <0 to 15 °C	Making use of solar gains, protection against heat loss
Tundra climate	Dry, monthly average temperatures from -20 to just over 0 °C	Protection against heat loss

Overhanging roof (protection from precipitation)
Walls with many openings (optimum cross ventilation)

Overhanging roof (protection from precipitation)
Overhang also protects from solar radiation
Walls with many openings (optimum cross ventilation)

Overhanging roof (solar protection, protection from heavy rains in the rainy season)
Solid construction (storage mass)
Facade closed towards main wind direction

Solid construction (storage mass)
Small openings (avoids overheating)

Solid construction (storage mass)
Flexible sun screening
Openings optimized for solar gain

Pitched roof (reduces snow loading)
Solid construction (storage mass)
Openings optimized to reduce heat loss

Solid construction (storage mass)
Openings optimized to reduce heat loss
Openings can be closed to minimize heat loss

Gando secondary school, Burkina Faso, 2001, 2008, Francis Kéré.

The predominant climate in Burkina Faso is that of tropical savanna. Dry seasons alternate with rainy seasons and the average temperatures throughout the year are between 25 and 30 °C.

The school building is adapted to this climate. The cantilevering roof construction protects the interior from direct solar radiation and the solid brick construction of the walls protects it against heavy rain. There are windows with folding shutters on both sides of the classrooms to provide optimum cross-ventilation as well as protection against the sun. Heated air is discharged via the roof space. Suspended cloths limit the heat radiating from the tin roof down into the room. The building does not feature any active cooling systems. Primarily locally available materials were used for the construction.

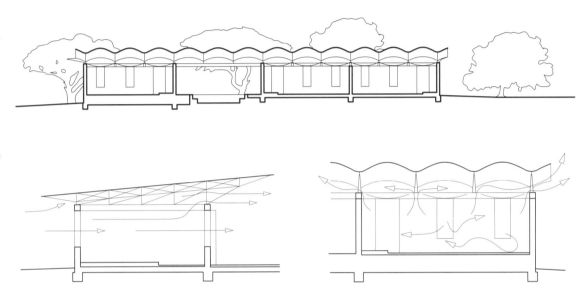

MESOCLIMATE

The specific climate of a region or conurbation, a valley, or other succinct geographic entity is referred to as a mesoclimate. In this context, the transition to the higher level macroclimate and the lower level microclimate is gradual. The mesoclimate of a region is affected by local parameters such as vegetation and surface cover, surface form, regional precipitation quantities, as well as human activity. In many cases these parameters have led to the development of building typologies that form the shape of buildings and unify urban development patterns.

PRECIPITATION
The quantity of precipitation in the form of rain or snow depends largely on topography and wind direction. In Germany, most precipitation falls with westerly and northwesterly winds. Clouds pile up against the county's low mountain ranges and the Alpine foothills, resulting in precipitation. This means that precipitation is higher particularly to the west of these areas of high ground. Depending on the region, precipitation in Germany is between 500 and 2,000 mm per square meter per year.

WIND
Different wind systems and their different origins can have an impact on the mesoclimate of a region.
In coastal areas by the sea or near large inland lakes, the so-called land-sea wind is a common feature. In temperate latitudes this wind system extends along the coast in swaths 10 to 20 km wide. Along hot coastlines it can extend up to 100 km inland and cause temperature differentials of up to 10 K. In summer, buildings near the coast are cooled by the incoming sea air and in winter the constant exposure to cold air results in an increased heating requirement.
In ranges of low mountains or high hills, two different wind systems occur in combination. On the one hand there is slope wind circulation and, on the other hand, interacting with the Alpine foothills, the mountain-valley wind system. Slope winds are created during daytime as the mountain slopes facing the sun warm up. The heated air carries water vapor, which during the course of the day leads to formation of convective clouds above the hills or mountaintops. Convective clouds reduce the intensity of solar radiation in the higher regions, particularly during summer, and hence reduce the usefulness of solar thermal or photovoltaic installations. Mountain-valley winds affect the temperature in valleys. In Alpine valleys, the temperature fluctuation between day and night is approximately twice as much as that in the Alpine foothills. This effect can be largely reduced by the use of more thermal mass in buildings of solid construction.

GLOBAL RADIATION
The term *global radiation* refers to the solar radiation that hits the earth's surface. A distinction is made between direct and diffuse radiation, with the latter referring to direct radiation scattered and reflected from clouds, water, and dust particles in the atmosphere. Global radiation is a source of solar heat and plays a part in heating up buildings. The intensity of global radiation and the proportion of diffuse radiation in a location are subject to daily and annual fluctuations. In addition, global radiation is affected by aspects of the weather, such as cloud formation and atmospheric opacity. In central Europe, global radiation on a cloudless summer day amounts to 900 W/m² on a horizontal surface. When there is light cloud cover, this value can rise to over 1,000 W/m² (diffuse radiation). When there is heavy cloud cover, these values can drop down to about 100 W/m². The annual total of global radiation in Germany is between 900 and 1,200 kWh/m² per annum. Annual global irradiation is increased as the angle of irradiation increases towards the equator.

1 Functional principles of localized wind systems (daytime).

2 Functional principles of localized wind systems (nighttime)

a Land-sea wind.
During the day, the land heats up faster and hotter than the water surface, warm air rises and draws in cool air from the water (onshore wind). During the night this effect is reversed, the land cools down faster, warm air above the water surface rises and draws in air from the land (offshore wind).

b Slope winds.
Warm air rises up the south-facing slope and draws in air from the base of the valley and the opposite shaded mountain slope. At nighttime this effect is reversed.

c Mountain-valley wind.
In the mountain-valley wind system, the relatively small air quantity in the mountain valleys heats up faster during daytime than the air in the Alpine foothills, air from the Alpine foothills flows up the valleys (valley wind) whereas at night, the air in the valleys cools down faster and air from the valleys drops into the foothills (mountain wind).

d Katabatic wind.
On slopes where the lower air layer is subject to pronounced cooling and the air slides into the valley, so-called katabatic winds develop. Typical examples are the winds coming from glaciers in the high mountains. Glacial winds always go downhill irrespective of the time of day.

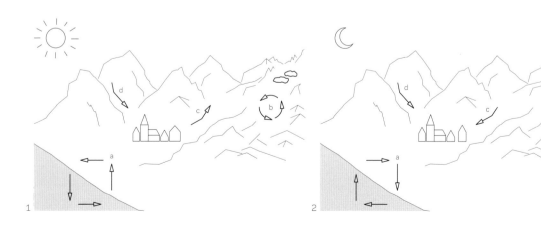

1   2

The distribution of global radiation in Germany is not uniform. There are regional differences which make it necessary to study the subject in greater detail. A general difference between north and south is discernible. Southern Germany in particular has the highest average annual global radiation.

3 Distribution of global radiation in Germany.

900-950 kWh/m²a
950-1,000 kWh/m²a
1,000-1,050 kWh/m²a
1,050-1,100 kWh/m²a
1,100-1,150 kWh/m²a
1,150-1,200 kWh/m²a

4 Distribution of direct and diffuse radiation in Germany per month.

5 Average amounts of precipitation and distribution in Germany (1971-2000).

Under 500 mm/a
500-650 mm/a
650-700 mm/a
700-800 mm/a
800-1,000 mm/a
1,000-1,800 mm/a
Over 1,800 mm/a

6 Distribution of precipitation in Germany per month.

3

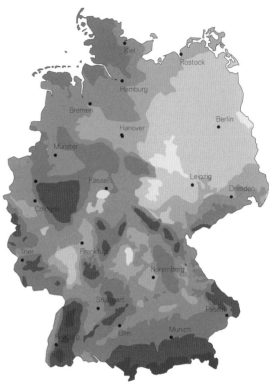

5

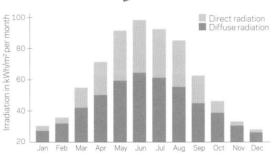

4

6

7 Adaptation of building construction to local weather conditions:

a Arcades in Bologna protect from the sun and precipitation, and they create an airy and open urban environment.

b Houses in Goslar, Germany, have slate shingles covering the sides exposed to the weather in order to protect them against driving rain, conveying a closed and protected impression.

7a

b

MICROCLIMATE / URBAN CLIMATE

The microclimate refers to the climate of a specific local area such as a building complex. The microclimate of the immediate environment of buildings has a direct influence on the thermal conditions in these buildings. ⌐2 Usable global radiation can be reduced by shading from the building itself or other objects, but it can also be enhanced through reflection from other surfaces in the proximity. ⌐3 Design information about usable solar radiation can be obtained from sun position models and simulations. ⌐5 Water and green areas in the vicinity have a cooling effect due to evaporation. During summer wind has a cooling effect, but in winter constant exposure to cold air increases the heating requirements of a building.

URBAN CLIMATE

The inner urban microclimate is affected by a number of factors. ⌐4 Overheating increases with increasing density and ground sealing. Compared to the surrounding open countryside, temperatures in cities can increase by 1–3 K ⌐1, in extreme cases by up to 10 K. On the one hand, this heat increase results from man-made generation of heat (emissions from heating systems, industry, and traffic) and from increased absorption of long-wave radiation due to denser cloud cover. On the other hand the heat is not carried off to the same degree, as wind forces are reduced by up to 30%, and the proportion of green space (approx. 5–20% of urban area) is reduced, thus limiting the cooling evaporation effect. Although this heat gain means that the heating requirement of buildings in urban areas is reduced in winter, it leads to an unpleasant increase in temperature and humidity during summer. On the level of urban design, these problems can be addressed by maintaining / creating connected green spaces of the largest possible size, keeping air flow corridors clear and utilizing slope wind situations for nighttime cooling. Surfaces should be as reflective and absorb as little solar radiation as possible (light surfaces).

North; wind direction ⟶

	Open position (reference value)	Sheltered position	Basin; cold air pockets	South-facing slope	Hilltop, mountaintop
Ambient temperature differential	+/-0 °C	No data	-3 K	+2 K	-1 K
Heat losses	+/-0%	No data	+ 25%	-17%	+10%
Heat loss through wind	+/-0%	-50%	No data	+/-0%	+100%

2

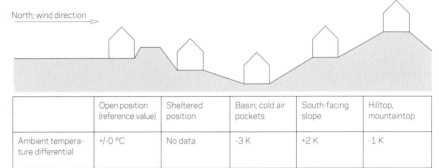

3

Air temperature near the ground	+1 K	to	+3 K
Global radiation	-15%	to	-20%
Ultraviolet radiation	-10%	to	-30%
Duration of sunshine	-5%	to	-15%
Relative humidity near the ground	-2%	to	-10%
Precipitation quantity	+5%	to	+10%
Number of rain days		approx	+10%
Cloud cover	+5%	to	+10%
Wind speed	-10%	to	-30%
Number of days with fog and poor visibility	+50%	to	+100%
Aerosol content in the air		to	+1,000%

1

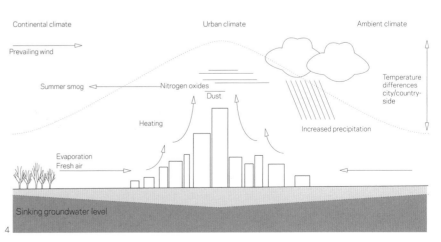

4

1 Differences between the urban climate and the climate of the sur-rounding areas.

2 Effects of position and exposure to wind on heat loss of a building.

3 Microclimate factors:
a Evaporation cooling from trees and green areas
b Reflection of sunlight from water surfaces
c Adiabatic cooling through water surfaces
d Self-shadowing
e Facade greening
f Shadowing by other objects
g Heat radiating from sealed sur-faces

4 Functional principles of urban climate.

5 Simulation of the sun's move-ment for the elevations of a building in a gap.
a Urban situation
b Solar irradiation on the floor lev-els of different stories at the sum-mer and winter solstice respec-tively and also at the spring and autumn equinoxes (north orienta-tion)
c Solar irradiation hitting the southern, eastern, and northern facades on 21 March
d Solar irradiation hitting the southern, eastern, and northern facades on 21 June
e Solar irradiation hitting the southern, eastern, and northern facades on 21 December

6 Microclimate mapping and ther-mal assessment of an inner urban open space. Thermal classification from cold to warm.

7 Design proposals derived from thermal mapping:
a Medium-high row of trees to avoid heat being radiated off
b Sun space with water-bound surface
c Trees for shading
d Trees as wind protection
e Facade greening for cooling and reducing overheating

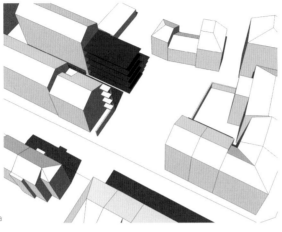

5a

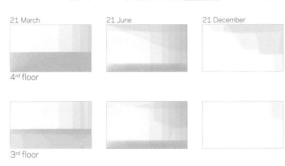

21 March   21 June   21 December

4ʳᵈ floor

3ʳᵈ floor

2ⁿᵈ floor

b 1ˢᵗ floor

MICROCLIMATE SIMULATION

Different analytical methods are available for determining the microclimate situation of a locality for the purpose of building design. In a microclimate simulation, all elements of the urban environment (buildings and vegetation) are entered into a software program and assigned surface characteristics (e.g. reflection and absorption). The sur-face temperatures during the simulation period (as a rule, 24–48 hrs.) are determined by taking into account locally measured meteorological data, such as air temperature, wind speed, wind direction, humidity, and global radiation. These temperatures are recorded in a microclimate map.
➔ 6 Microclimate simulation makes it possible to illus-trate criteria that have a direct effect on the design of a building. Examples are the albedo, the average radiation temperature of a facade, and the effect of vegetation on wind speed, humidity, and air temperature. ➔ 7

c South   North

d South   North

e South   North

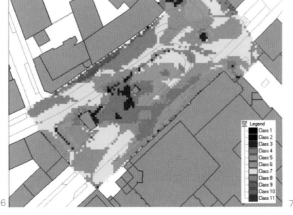

6

Legend
Class 1
Class 2
Class 3
Class 4
Class 5
Class 6
Class 7
Class 8
Class 9
Class 10
Class 11

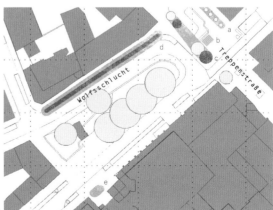

7

URBAN STRUCTURE

Urban structures of an existing or planned development use up natural space, and urban sprawl increases the need for transport; in addition they have a direct effect on the energy balance of individual buildings. The density and type of building development is determined by climate conditions, historic development, and details specified in official development plans.

RATIO OF A TO V_e

The ratio of the surface area of the envelope of a building (A) to its built volume (V_e) is a measure of the compactness of a building. The lower the A/V_e ratio is, the more compact the building, which means that less energy is lost through the building envelope relative to its volume. → p. 68 The achievable A/V_e ratio and the resulting transmission heat loss (H_T') of the buildings, i.e. the energy lost through their envelope, depend on the urban structure and the planned building typology. → 2 Highly structured building shapes are less compact. Compactness is higher for larger buildings of the same shape. There can be a conflict between the compactness of a building and its ability to be lit by natural light. Where this has to be supplemented with artificial light, the result is higher energy consumption.

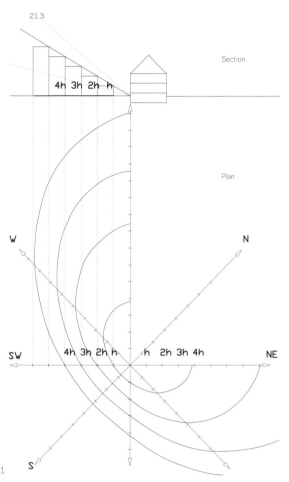

1

1 Distance to adjacent buildings required in order to ensure at least 2 hours of solar radiation on the eastern, southern, and western facades at equinox in relation to the height of the surrounding buildings. The distances vary for each of the cardinal directions due to the sun's movement during the course of the day. The smallest possible distance that still allows full solar irradiation of the facade is towards the south.

2 A/V ratio for different building types (building envelope and building types in accordance with EnEV 2009). Size per unit with flat roof: 12 m x 8 m x 3 m. The envelope surface per unit, through which heat can escape, is in large multi-family dwellings less than half that of single-family dwellings. At the same time however, the roof area available for solar radiation is also reduced.

		Detached single-family dwelling	Semi-detached single-family dwelling	Terraced house (row house)	Small detached multi-family dwelling	Large detached multi-family dwelling	Large semi-detached multi-family dwelling	Large multi-family dwelling at end of terrace
Number of units	[-]	1	1	1	4	8	8	8
Total volume	[m³]	288	288	288	1.152	2.304	2.304	2.304
A/V_e ratio	[1/m]	1,08	0,96	0,83	0,63	0,46	0,40	0,33
ø Area of envelope per unit	[m²]	312	276	240	180	132	114	96
ø Area of unit exposed to external air	[m²]	216	180	144	132	108	90	72
ø Area of unit in the ground	[m²]	96	96	96	48	24	24	24
Roof area of unit available for solar energy collection	[m²]	96	96	96	48	24	24	24
Ratio of building area exposed to ground to building area exposed to external air	[-]	1:2,3	1:1,9	1:1,5	1:2,8	1:4,5	1:3,8	1:3,0

2

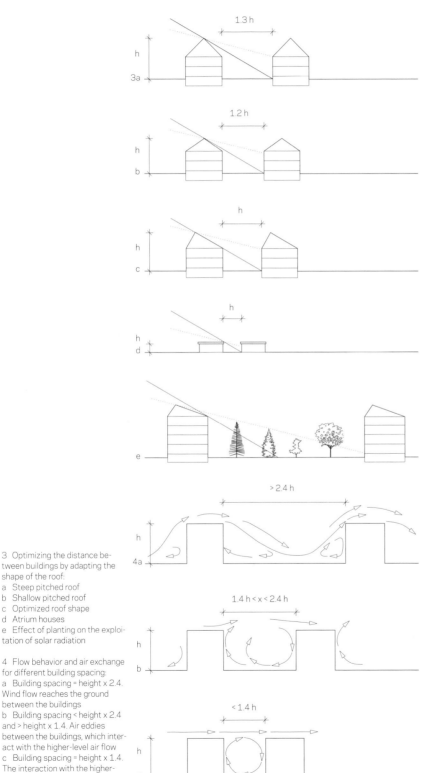

3 Optimizing the distance between buildings by adapting the shape of the roof:
a Steep pitched roof
b Shallow pitched roof
c Optimized roof shape
d Atrium houses
e Effect of planting on the exploitation of solar radiation

4 Flow behavior and air exchange for different building spacing:
a Building spacing = height x 2.4. Wind flow reaches the ground between the buildings
b Building spacing < height x 2.4 and > height x 1.4. Air eddies between the buildings, which interact with the higher-level air flow
c Building spacing = height x 1.4. The interaction with the higher-level air flow is limited

SPACING, LIGHTING, AND VENTILATION

The spacing between buildings and their height affect the amount of solar radiation reaching their facades and hence influence passive solar gain, and determine whether integrated solar technology can be installed. Depending on the orientation of buildings, it is possible to space them differently in relation to neighboring buildings. The design should aim for a situation in which each external wall facing east, south, and west has at least 2 hours of sunshine on the equinoxes of 21 March and 21 September (2-hour rule). ⤵ 1 Where solar technology is integrated into the facade of a building, the solar irradiation period should be significantly longer (> 4 hours). The shape of the roof also has an effect on the shading of a building. By adapting the shape of a roof it is possible to optimize the spacing between buildings and thus achieve greater density. ⤵ 3 On the other hand, urban designs based purely on the optimization of the orientation and spacing between buildings may lead to a monotonous appearance. This gives some indication of how design and energy efficiency affect each other. It follows that an optimal design can only be achieved when all aspects of sustainable urban design are taken into account.

The solar irradiation of buildings is also affected by the height and type of planting in the vicinity. ⤵ 3 Deciduous trees, which shed their leaves in winter, do not obstruct solar irradiation, and can therefore be used to raise the temperature of buildings, whereas deciduous trees in leaf overshadow buildings and thereby prevent undesirable overheating. Coniferous trees overshadow buildings throughout the year and therefore reduce useful solar irradiation during winter. Taking into account the height of trees, spacing between them and buildings should be determined by following the same rules as those for spacing between buildings. Special consideration needs to be applied to the refurbishment of existing buildings where adjacent vegetation may have grown high and dense. The height of and distance between buildings also affect the air changes in the urban environment. Where buildings are sited too closely together, the remaining open space is not sufficiently ventilated, the pressure differential between the different elevations of a building is reduced and hence the options for ensuring adequate natural ventilation in the building are limited. The optimum level of ventilation of the space between buildings can be achieved where the spacing between buildings is greater than three times their height. ⤵ 4 It follows that in order to achieve good ventilation, greater distances between buildings are needed than those required for solar irradiation of the facade; this may cause a conflict where space is at a premium.

POTENTIAL FOR USING GEOTHERMAL AND SOLAR ENERGY

The degree of utilization of a building plot is defined by its building coverage ratio (BCR) and floor area ratio (FAR). These ratios can be used to roughly calculate the areas available for utilizing solar and geothermal energy. In order to make use of solar energy, solar installations on the roof (photovoltaic or solar thermal panels) promise the greatest yield, as there is likely to be the least shading here.

If the intention is to cover all the electricity required for air conditioning, computing, and domestic equipment through a photovoltaic installation integrated in the roof → p. 107, then the number of stories should be between two and five (maximum). For buildings higher than five stories, the ratio of available roof area to usable floor area (and hence the number of users) is too low. → **1**

To cover the heat requirement for generating hot water with the help of a solar thermal installation on the roof, the maximum number of stories can be between 10 and 20 (maximum). → **2** The difference in the requirements for a photovoltaic installation is partly due to the higher efficiency of solar thermal systems and partly due to the lower energy consumed in the preparation of domestic hot water compared to the overall power requirements of a building. Where such installations are integrated in the facade, particularly with high buildings, there is more potentially suitable area available and it is also closer to the user (pipe losses of solar thermal energy are minimized). Moreover, vertical surfaces can make better use of solar radiation in winter, when the demand for hot water is usually higher.

If the intention is to use geothermal energy (from near-surface sources in combination with a heat pump) to cover a large part of the heating and cooling energy required by a building, the optimum floor area ratio for residential buildings is 3 to 5 (maximum). → **3** The potential for utilizing geothermal energy is largely dependent on the type of soil as well as the density and type of building development. Particularly suitable is a loam soil with a high water content; dry sandy soils are less suitable. To assess the potential accurately, it is recommended that a soil investigation report be commissioned. In contrast to geothermal collectors, the installation in Germany of geothermal ground loops is subject to official approval. It is also possible to use groundwater as a heat source, which involves the installation of a well for pumping the water to the surface and an infiltration sink for discharging it. This method is economical where the groundwater is a maximum of 20 meters below the surface. Using the groundwater for heating purposes in Germany is subject to approval by the local lower administrative water authority.

AVAILABLE UTILITIES

Building sites may be serviced by a number of different utilities for their energy requirements. In addition to electrical energy (national electricity grid), these are usually natural gas or district heating. While the energy from electricity and natural gas is entirely or mostly derived from non-regenerative sources, 84% of district and local heating in Germany is provided by combined heat and power (CHP) stations. The energy provider will give information about the exact composition of the production. Where district heating systems are in existence, connection is often mandatory. In certain circumstances it is possible to apply for dispensation, for example where the existing or planned building has an emission-free heating system (e.g. a heat pump), or where the total energy requirement is so small that connection would not be economically viable. This can be the case in a highly insulated building (e.g. built to Passivhaus standard).

1 Potential of photovoltaic installations for covering the electricity requirements of different urban densities (defined by building coverage ratio and floor area ratio). Proportional contribution of roof-mounted photovoltaic installations to the electricity requirement of residential and office buildings.

2 Potential of solar-thermal installations to contribute towards the supply of domestic hot water for a number of different urban densities (defined by plot ratio and floor space index). Proportional contribution of roof-mounted solar-thermal installations to the domestic hot water requirement of residential buildings.

3 Potential of geothermal installations for heating and cooling of buildings for a number of different urban densities (defined by building coverage ratio and floor area ratio). Proportional contribution of geothermal installations (geothermal probes) to the heating and cooling requirement of residential and office buildings.

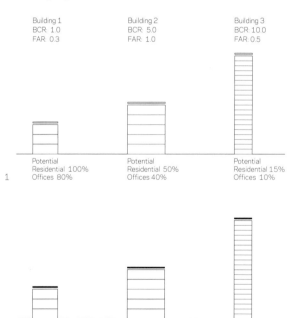

1

Building 1
BCR: 1.0
FAR: 0.3

Building 2
BCR: 5.0
FAR: 1.0

Building 3
BCR: 10.0
FAR: 0.5

Potential
Residential 100%
Offices 80%

Potential
Residential 50%
Offices 40%

Potential
Residential 15%
Offices 10%

2

Potential 100%

Potential 100%

Potential 40%

3

Potential 100%

Potential 50%

Potential 25%

EXTERNAL HEAT SOURCES AND HEAT SINKS

A wide variety of energy sources can be used for heating and cooling buildings; some of these are available directly on the building or site (ground, groundwater, solar radiation, external air, surface water, exhaust air, wastewater, wind) and some have to be supplied to the building via distribution networks or other infrastructure installations (natural gas, mineral oil, wood pellets, wood chips, plant oil, biogas, mains electricity, and district or local heating). Depending on the energy sources and utilities available, energy can be used in different forms (electrical or thermal). In addition, energy sources are available at different temperature levels, which in turn affects the type of heating system of the building.

4 Illustration of external heat sources and heat sinks (absence of heat; coldness), their possible forms of supply, forms of energy (electrical or thermal), and temperature level, if applicable.

Energy sources	Form of supply	Electrical	Thermal	Temperature level	Note
Natural gas Heating oil Wood pellets Wood chips Plant oil Biogas	Piped/delivered/heating boiler		x	60-90 °C	Steam boiler >100 °C
	Piped/delivered/CHP station (combined heat and power station)	x	x	70-90 °C	Approx. 20-40% electrical and 60-80% thermal, depending on size
Mains current	Power connection	x		-	
District heating (large or small scale)	Transfer station		x	90-130 °C	Low voltage 220 V
District cooling (large or small scale)			x	Approx. 5 °C	
Ground	Geothermal probes		x	12 °C	50-100 W/m probe, requires approval
	Ground loops		x	12 °C	10-35 W/m² ground loop
Groundwater	Geothermal probes		x	10-12 °C	Reversal of functions in summer and winter
Solar radiation	Flat plate collector (air)		x	Up to 80 °C	At least 10 °C above external temperature
	Flat plate collector (water)		x	80 °C	
	Vacuum tube collectors (water)		x	150 °C	Also suitable for process heat
	Photovoltaics	x		-	50-150 W_p/m²
External air	Heat exchanger		x	= External temperature	
Surface water			x	0-25 °C	Subject to seasonal variation
Exhaust air			x	20-24 °C	Temperature of industrial exhaust air can be higher
Waste water			x	15-23 °C	At least 25-50 residential units should be connected
Rainwater			x	= External temperature	Can be used for adiabatic cooling
Wind	Small wind turbine	x		-	Up to 100 kW per system

4

LEGAL BACKGROUND

The construction of buildings is subject to a number of regulatory provisions. The requirements for the characteristics of the building envelope and services installations in new and existing buildings in Europe is defined in European Directive 2002/91/EC. European directives are implemented in each member country's legislation (examples are the Energy Conservation Regulations in Germany and other countries, and the Energy Certification Providing Act in Austria). The German standard will be examined below as an example. Requirements also exist regarding the proportion of renewable energy used for the energy supply of buildings, such as the Renewable Energies Heat Act (REHA). In addition to regulatory provisions, there are a number of building energy standards, some of which are used as a requirement for obtaining public funding (e.g. KfW Effizienzhaus from the Credit Institute for Reconstruction) or for identifying particularly high energy standards (e.g. Passivhaus). Some of these more exacting standards apply their own calculation methods. ⌲ 2 - p. 39

ENERGY CONSERVATION REGULATIONS (ENEV) 2009

In Germany, the Energy Conservation Regulations define the requirements for the quality of the building envelope, heating and ventilation systems, and domestic hot water services. The EnEV applies to residential as well as non-residential buildings, both new and existing, with a usable area of more than 50 m².

The main requirement for newly designed residential buildings limits the annual primary energy requirement Q_P and the secondary requirement limits the specific transmission heat loss (H_T'), which refers to the average heat permeability of the building envelope. Two calculation methods are permitted for determining the annual primary energy requirement (in accordance with DIN EN 832 or DIN V 18599). Since EnEV 2009, the limits of the primary energy requirement and specific transmission heat loss are no longer determined by the A/V_e ratio, a change which allows greater freedom in the architectural design. Now the limits are determined with respect to reference buildings and building typology (detached, semi-detached, extension, other residential buildings).

In newly designed non-residential buildings the annual primary energy requirement Q_P is limited. In addition, average thermal transmittance values (U-values) have to be achieved for the components of the building envelope. In non-residential buildings, the energy requirement for lighting and air conditioning is also entered into the calculation. For these buildings the calculation method pursuant to DIN V 18599 is mandatory and the limits are determined with respect to appropriate reference buildings.

In addition to thermal insulation in winter, it is also necessary to provide proof of protection against overheating in summer, irrespective of the building typology. While the envelopes of buildings must be impermeable to air, the minimum air changes to comply with hygiene requirements have to be ensured (mechanical ventilation may be required). This minimizes ventilation heat loss through gaps in the building fabric. Thermal bridges must be avoided in accordance with the codes of practice. In row-house developments the party walls between buildings must be insulated where it is not certain that a neighboring building will be erected.

RENEWABLE ENERGY SOURCES ACT (RESA)

The Renewable Energy Sources Act obliges utility network operators to purchase electricity from renewable energy sources at a fixed price that is higher than the market price. The network operators pass on the additional cost to their end customers. This procedure supports the use of renewable sources of power production, which is to be increased to 30% of overall production by 2020 (16% in 2009). For building owners and architects, the generation of electric power with photovoltaic panels is of particular interest. The tariffs depend on the output of the installation. Systems with up to 30 kWp (150–250 m² of panel surface) are considered small, and it is mandatory for energy providers to connect these systems. Payment can be received both for feeding electricity into the network and for own use.

RENEWABLE ENERGIES HEAT ACT (REHA)

The Renewable Energies Heat Act is intended to increase the proportion of renewable energy used for heating, hot water, cooling, and process heat in buildings to 14% by 2020 (8.4% in 2009). To help achieve this goal, a minimum proportion of the heat required for heating a building must come from renewable sources. The actual amount depends on the type of renewable energy source. ⌲ 2
The act permits different measures to be used for meeting its requirements. As an alternative for complying with the proportion of renewable energy sources, it is also possible to meet the requirements of the act by achieving an improvement of 15% on the EnEV requirement. This is of particular interest for buildings constructed to Passivhaus standard, in which the residual heat requirement is covered by electric appliances. The REHA only applies to new buildings.

1 Calculation of final and primary energy requirement for heating and hot water, taking into account gains and losses (1 Data for average district heating with a combined CHP component of 70%).

Q_P Primary energy requirement
Q_F Final energy requirement for heating
Q_W Final energy requirement for hot water
O_g Losses incurred by generating companies
Q_s Storage losses
Q_d Distribution losses
Q_h Heat requirement for space heating
$Q_{c,e}$ Control losses
Q_T Heat losses via transmission
Q_V Ventilation heat losses
Q_i Internal heat gains
Q_s Solar heat gains

2 Minimum percentage and required conversion of renewable energies by energy source (REHA). The REHA requirements are met with one of the combinations of energy source and technology shown.

3 Regulations for upgrading existing buildings (residential and non-residential) and plant, and minimum requirements for building components following refurbishment in accordance with EnEV 2009.

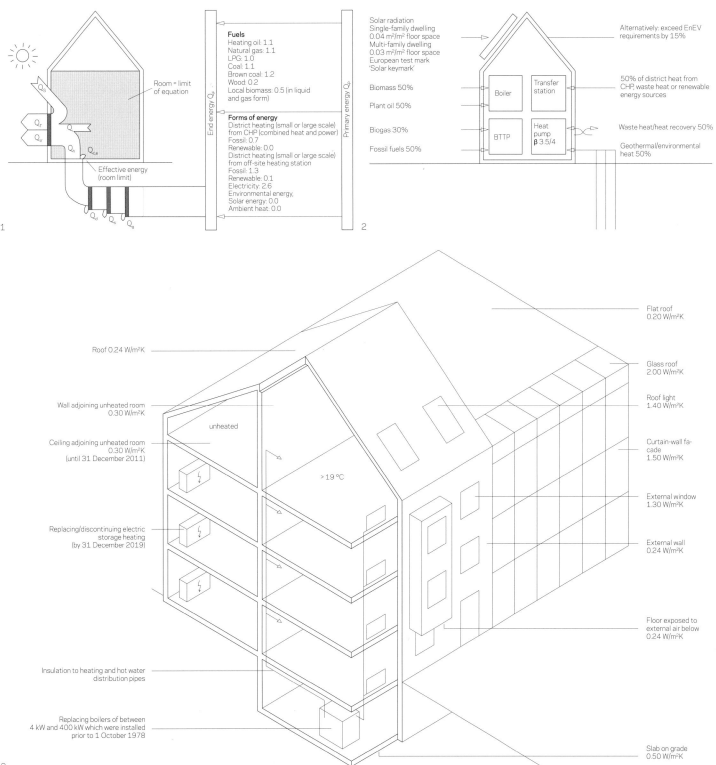

1

Fuels
Heating oil: 1.1
Natural gas: 1.1
LPG: 1.0
Coal: 1.1
Brown coal: 1.2
Wood: 0.2
Local biomass: 0.5 (in liquid and gas form)

Forms of energy
District heating (small or large scale) from CHP (combined heat and power)
Fossil: 0.7
Renewable: 0.0
District heating (small or large scale) from off-site heating station
Fossil: 1.3
Renewable: 0.1
Electricity: 2.6
Environmental energy,
Solar energy: 0.0
Ambient heat: 0.0

Room = limit of equation

Effective energy (room limit)

End energy Q_e

Primary energy Q_p

Q_S
Q_T
Q
Q_V
Q_n $Q_{c,B}$
Q_d Q_s Q_g

2

Solar radiation
Single-family dwelling
0.04 m²/m² floor space
Multi-family dwelling
0.03 m²/m² floor space
European test mark 'Solar keymark'

Biomass 50%

Plant oil 50%

Biogas 30%

Fossil fuels 50%

Boiler

Transfer station

BTTP

Heat pump
β 3.5/4

Alternatively: exceed EnEV requirements by 15%

50% of district heat from CHP, waste heat or renewable energy sources

Waste heat/heat recovery 50%

Geothermal/environmental heat 50%

3

Roof 0.24 W/m²K

Wall adjoining unheated room
0.30 W/m²K

unheated

Ceiling adjoining unheated room
0.30 W/m²K
(until 31 December 2011)

> 19 °C

Replacing/discontinuing electric storage heating
(by 31 December 2019)

Insulation to heating and hot water distribution pipes

Replacing boilers of between 4 kW and 400 kW which were installed prior to 1 October 1978

Flat roof
0.20 W/m²K

Glass roof
2.00 W/m²K

Roof light
1.40 W/m²K

Curtain-wall façade
1.50 W/m²K

External window
1.30 W/m²K

External wall
0.24 W/m²K

Floor exposed to external air below
0.24 W/m²K

Slab on grade
0.50 W/m²K

ENEV 2009 REQUIREMENTS FOR EXISTING BUILDINGS

The provisions of the 2009 Energy Conservation Regulations also apply to modifications, extensions, and refurbishment of existing buildings. Where existing buildings (residential and non-residential) are modified, the external building components involved have to comply with certain values for thermal transmittance. ⟍ 3 - p. 37 This applies where the construction affects more than 10% of the surface of building components. Alternatively, the requirements are met when the refurbished building does not exceed the primary energy requirement and specific transmission loss (residential building), or the primary energy requirement and the requirements regarding components of the building envelope (non-residential building) by more than 40% compared to a new building of equal size (reference building).

EXTENSIONS

Where existing buildings (either residential or non-residential) are extended, coherent extensions with usable areas of between 15 and 50 m² are subject to the simplified provisions for small buildings, which means that the thermal transmittance values of the envelope of the extension building must be within certain limits. ⟍ 3 - p. 37 If the extensions are larger than 50 m² usable area, they have to be treated as new buildings.

INSTALLATIONS

For existing buildings, whether extensions or conversions, EnEV 2009 stipulates a number of upgrade installations and/or the removal of certain installations.
Heating boilers with an output of between 4 and 400 kW that were installed prior to 1 October 1978 and are not low-temperature or condensing boilers may no longer be operated. Exempt from this rule are non-proprietary boilers (special designs) and boilers used exclusively for heating domestic hot water.
In buildings with
- more than 5 apartments (residential buildings) or
- more than 500 m² heated usable area (non-residential buildings);
- for which the building application was made prior to 31 December 1994;
- and which do not comply with the provisions of the 1994 Thermal Insulation Ordinance;
it is not permitted to operate electric storage heating systems where these are the sole heat source and have a heat output of more than 20 W/m². In all other buildings, electric storage heating systems that were installed prior to 1 January 1990 may no longer be operated after 31 December 2019. Electric storage heating systems that were installed after 1 January 1990 may not be operated beyond a period of 30 years.

BUILDING COMPONENTS

In existing buildings, the energy performance of certain building components has to be improved, irrespective of whether they are part of an extension or a conversion. Domestic hot water and heating distribution pipes without insulation that run through unheated rooms must be insulated. ⟍ 3 - p. 37
Floor slabs between rooms and unheated loft spaces that are accessible but cannot be walked on have to be insulated to reach a thermal transmittance of 0.24 W/m²*K. Alternatively, it is possible to appropriately insulate the roof above.
From 31 December 2011, this regulation will also apply to floors between rooms and unheated loft spaces that can be walked on (e.g. drying lofts, attics).

Building owners must carry out these measures within the specified time frames. In owner-occupied dwellings with up to two apartments, the above-mentioned upgrades do not have to be implemented until there is a change in ownership. None of these measures have to be implemented unless commercial amortization is possible within a reasonable period of time. The only exception is the replacement of old heating boilers, which have to be replaced in all cases. Again, these measures are checked by district certified chimney sweeps.

FUTURE DEVELOPMENTS

The rapid succession of new legislation in recent years has led to an improvement in the energy performance of new buildings. However, owing to the low rate of new construction and the long-term action of changes in architectural design, these improvements take a long time to affect the total building stock. For this reason, architects need to be aware of future developments so that they do not design buildings with an energy performance that is outdated within a few years. The following basic developments are likely: with the coming EnEV amendments (probably in 2012 to 2015), the requirements for building envelopes will reach those of the Passivhaus standard. Thereafter, further improvements will concern services installations. According to current planning, the European Union is set to stipulate that, by 2020, all buildings must generate nearly all the energy used in them. In parallel, there will be a shift of emphasis away from energy consumption towards limiting emissions (CO_2 and others). Furthermore, as energy consumption is reduced during the use-phase of buildings, consideration of the energy used in the production, maintenance, and disposal of building components will become more prominent (life cycle considerations).

1 Development of regulations and laws in Germany over the last forty years, and future development. The resulting development of the heating requirement/energy balance of buildings. Current planning horizon.

2 Different building energy standards with level of comparison and limits.

According to plannning of the European Union, buildings will generate the energy they require for heating and hot water themselves from about 2020, i.e. the energy balance of new buildings will become positive (plus-energy house). When taking into account the stock of existing buildings and assuming a refurbishment rate of 2% and a demolition rate of 1% per annum, the effect on the total building stock is delayed by approx. 30 years (dotted line). Depending on the type of building, buildings designed today have different service life expectations, which significantly exceed this period.

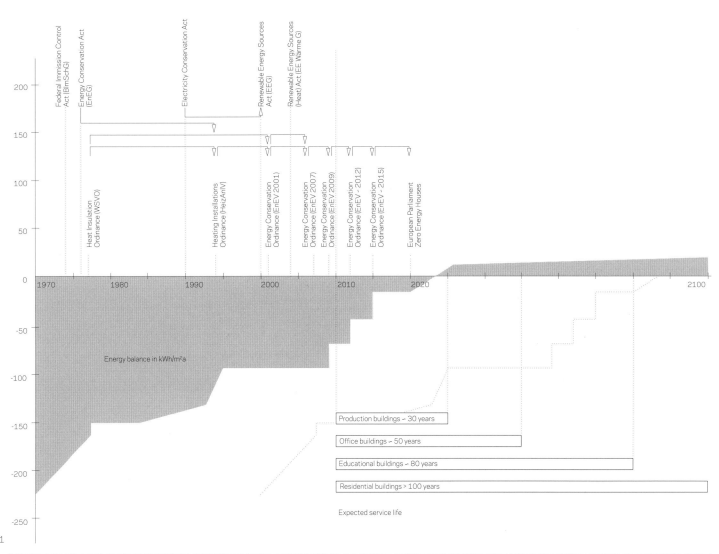

1

Standard		Level of comparison	Limits	Notes
EnEV: residential buildings	D	Primary energy requirement for heating, ventilation and domestic hot water	Determined in relation to reference buildings	Legal minimum requirement for new construction in Germany (limit for refurbishment 40% higher)
EnEV: non-residential buildings	D	Primary energy requirement for heating, ventilation, drinking water, refrigeration and lighting		
KfW Effizienzhaus 40, 55, 70	D	Primary energy parameter of reference building thermal insulation standard pursuant to EnEV	Q_p to be lower by 60%	45%	30% H_T' to be lower by 45%	30%	15%	Requirement for public funding for new buildings, as of October 2010
KfW Effizienzhaus 40, 55, 70	D		Q_p to be lower by 45%	30%	15%	0%	-15% H_T' to be lower by 30%	15%	0%	-15%	-30%	Requirement for public funding for existing buildings, as of October 2010
Minergie house	CH	Weighted energy parameter (end energy): heating, ventilation, domestic hot water, air conditioning	Max. 38 kWh/m²a, 90% of limit according to Swiss SIA standard 3800| 1:2009	Further requirements: e.g. building envelope, mechanical ventilation, upper limit of construction costs
Minergie-Plus house	CH	Weighted energy parameter (end energy): heating, ventilation, domestic hot water, air conditioning	Max. 30 kWh/m²a, 60% of limit according to Swiss SIA standard 3800| 1:2009	Further requirements: air tightness, max. heating output, heat requirement for space heating, electricity requirement
Passivhaus	D	Heat requirement for space heating; primary energy requirement for heating, ventilation and domestic hot water	Max. 40 kWh/m²a	Total primary energy requirement, including electricity max. 120 kWh/m²a
Net Zero Energy House	D	Primary energy requirement for heating, ventilation and domestic hot water	Zero annual balance achieved through credits (for micro-generation)	Credits usually from photovoltaics or combined heat and power generation
Plus Energy House	D	Primary energy requirement for heating, ventilation and domestic hot water	Positive energy balance achieved through credits/ electricity network feed-in	Credits usually from photovoltaics or combined heat and power generation

2

HEAT | COOL
INTERNAL FACTORS

INTEGRATED SPACE DESIGN

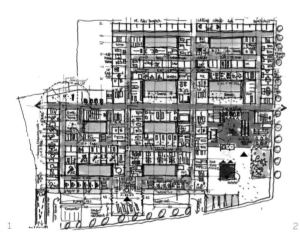

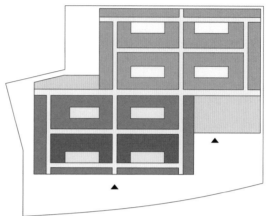

The internal factors of buildings with highly specialized uses, such as hospitals or certain production processes, are heavily affected by their functional requirements. Conversely, in other types of building projects there is more design freedom.

1 Sketch of hospital layout.

2 Usage zones.

For a long time, regulations and standards relating to energy conservation in construction dealt exclusively with the flow of heat between the interior and the exterior through the building components, i.e. with the building as a whole and its environment. In Europe this has led to a steady improvement of construction standards which, when properly applied, guarantee specified comfort levels inside buildings irrespective of wind and weather, and at the same time reduce the energy requirement in comparison with previous standards. Therefore, legal requirements for the construction of buildings deal with the question of whether the building envelope is sufficiently insulated in order to minimize heat loss during winter, and whether suitable and sufficient solar screening is installed in order to avoid overheating in summer. A few years ago, this strategy was amplified to include questions regarding technical services. This is aimed at improving efficiency and integrating renewable energy sources.

These processes are necessary for the development of construction, but by themselves are not sufficient. If energy-efficient construction is reduced to climate-adapted construction focusing on the relationship with the external environment only, other, equally important internal factors are not addressed. We are building ever more complex buildings for increasingly complex uses. The inner structures, and hence the requirements resulting from the building's uses, are becoming more and more important. Therefore, integrated design looks at both internal and external factors and tries to make use of synergies. This is particularly relevant for buildings designed to operate with a fraction of the previously required energy input, and which therefore react sensitively to changes in climate and structure or to extreme conditions.

SYSTEMIC APPROACH

Modern buildings are complex systems. A system is deemed "complex" when its individual parts cannot be considered in isolation and do not behave in a deterministic way. The behavior of a building with all its users is more akin to a dynamic process than it is to mechanical clockwork. In order to be able to describe the behavior of a building and its users interacting with it, one needs to take into account interactions between different areas and services. To this end, the building is subdivided into use zones with different energy requirements and characteristics.

The objective of subdividing a building into zones is not to balance the total energy and heat flows within the building. This could only be done at a stage when the detailed planning of the building is nearly completed and would therefore be too late to be of use for an integrated space design. Instead, this method is aimed at deepening the understanding of the building's fundamental functions and at analyzing the design. The zone model deals with qualitative characteristics and can provide answers to the following questions:
- Is the position of the zone in the building suitable for its use? Alternatively: have the uses been allocated to suitable and favorable areas of the building?
- Is it possible to benefit from synergies resulting from the interaction of zones?
- Is it possible to benefit from synergies resulting from the interaction of the services systems, such as heating/cooling, ventilation, and lighting?
- Are the specified requirements justified, or would it be reasonable to make reductions?

3 Organizational principle and size distribution.

4 Space allocation.

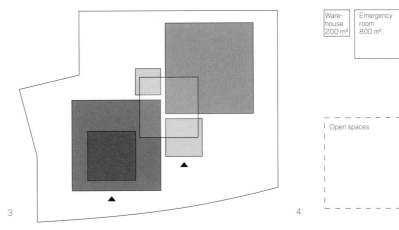

3

4

Example: Office
Offices can be organized in layouts ranging from cellular to open plan. These different layouts, together with their use patterns and occupation periods, have a big impact on the spatial organization as well as the energy requirement for ventilation and lighting.

Example: School Building
Changing half-day schools to all-day schools not only affects time arrangements but also the provision and layout of rooms in the building. The extension of functional and space requirements to include areas such as cafeterias, refectories, recuperation areas, and areas for sport activities and project groups with their specific needs, impacts on the profile of the energy requirement.

Likewise, conceptually based design decisions also have an impact on the energy requirement:
- the properties of selected building materials have an impact on the building's ability to store heat and absorb moisture, as well as on any potential emissions;
- the way materials are installed impacts on their ability and speed of response to heat exchange;
- the surface texture of components has an effect on thermal and acoustic qualities in the rooms;
- the color scheme affects the reflection or absorption of light.

SPACE ORGANIZATION

One of the core tasks of the architect is the subdivision, arrangement, and design of rooms. However, the analysis of spaces should not only be carried out under formal or aesthetic aspects, but should also take energy performance into consideration. In project development, every project starts with listing functional and space requirements. It is the first structured briefing upon which architects and engineers base their designs. With the help of these functional and space requirements, it is possible to determine the scope and type of the most important building functions without going as far as establishing physically-defined spaces. Usually, the functional and space requirement is produced in the form of a list of rooms with their sizes in square meters. In complex projects the space requirement of all the rooms is represented in one diagram, which shows sizes as well as relationships between functions (e.g. production processes). Once the functional and space requirements have been defined in greater detail and expanded by functional attributes, room data sheets (RDS) are created, sometimes also referred to as the specifications. This document includes the design results for the purposes of tendering and is also used later as documentation for operation of the building.

For his design, the architect uses the room data sheets as one of his sources of information—in particular the floor area and geometric arrangements of the different functional areas. He then arranges and allocates the various functional areas in the building and creates interesting spatial sequences and logical circulation systems, all of which is part of his original and creative work. Typical data included in the RDS are the room area, volume and clear height, surface finishes and materials, services installations, as well as furniture and fittings—in other words, more detailed input data beyond that of a description of the room function.

ENERGY CONCEPTS

The information compiled in this way can also be used for the development of energy concepts. For this purpose, the RDS data needs to be supplemented by energy-related information and effects. The type of material and surfaces, the type of construction and storage mass, the type of load-bearing structure and windows without lintels, construction details, and the routing of installations create structural conditions for all subsequent concepts, and therefore have to be approached holistically. The aim is to extend the design process and to include energy-related considerations—this may even result in the building gaining its own unique visual appearance based on these energy-related design elements.

But more important than the building fabric and physical properties are considerations about the organization and operation of the building. The spatial and time-based sequences of functions on their own do not determine the resulting energy requirement. Efficient spatial and services concepts and their dimensions can only be established on the basis of detailed information. This should be established in dialogue with the user, and it should then be analyzed with respect to its effect on energy consumption and the result fed back to the user. In this way it is often possible to achieve significant savings through a modification to outmoded user habits and an intelligent organization of uses with respect to space and time. This may make it possible to avoid technically complex and hence expensive solutions.

DEVELOPING ENERGY CONCEPTS

DIFFERENT APPROACHES

Buildings should provide their users with optimum conditions for living, working, learning, and recreation at all times. For the architect this means creating high quality rooms and a congenial environment. For the services engineer this means ensuring that conditions are comfortable in all situations, in short, that rooms have the appropriate amount of heating, cooling, lighting, ventilation, and electricity. While the architect defines rooms based on qualitative parameters, such as proportion and the quality of surfaces, the services engineer deals with quantitative parameters, such as floor area, volume, compactness, and storage mass. Both these professionals need to understand that they are talking about the same things, only in different ways. Both describe the properties of rooms with reference to envelope, volume, materials, and function. If a holistic approach is their shared aim, architects and services engineers need to find a common attitude and design method.

Since design quality and energy efficiency are not contradictory, the architect should check his design against energy-related criteria. At the concept stage it is possible to reduce the complexity of the design by taking into account the services engineer's considerations regarding heating, cooling, lighting, and ventilation. Typical questions at that stage would then be whether a room is ideally designed, acoustically and visually, for both winter and summer. In subsequent steps, a systematic analysis taking into account the interaction between the technical services and zones can be performed. The aim is to create optimum conditions and synergies both within the same zone and between different zones.

ZONING PROCEDURE

In the first step, rooms with the same function can be grouped together in a zone since we can assume that all these rooms have similar requirements. Depending on the activity, a specific temperature range is desirable, defined air changes need to take place, and an appropriate level of lighting must be provided. Depending on the position of rooms within the zone in respect to external factors (sun, distance to other buildings, vegetation, exposure to wind, etc.) and internal factors (conditions in adjoining zones), each zone must be further subdivided as they could otherwise tend to overheat or cool too much, in spite of their homogeneous functions. The result of zoning is a layout plan that includes use- and position-specific information and indicates the qualities, as well as the deficits, of the rooms.

DEFINITION OF USE PROFILES IN DIN V 18599

Since the introduction of DIN V 18599, the assessment of net and primary energy requirements of non-residential buildings, which is done as part of the statutory certification process, is based on a multi-zone model. The single-zone models, which were previously applicable, are no longer permitted except for residential buildings and non-residential buildings with just one predominant use. In addition to introducing multi-zone models, the scope of performance assessment was also expanded. For decades, these calculations were based on the energy required for heating buildings; this has now been extended to include energy needed for cooling, ventilating, and lighting a building. In Part 10, the standard refers to use profiles for various types of use. The need to subdivide non-residential buildings into different zones arises because different uses may have fundamentally different internal requirements.

In situations where the building project is commissioned by somebody who is not the future user, the design should be as flexible as possible to avoid the necessity of conversion work, or should provide for any anticipated changes in use. Therefore, the energy concept should not only be efficient but also robust when it comes to changes in use.

It should also be remembered that the furniture and fittings in rooms can change. After many years of continual increase in internal heat output through electrical equipment, the trend is now towards greater energy efficiency (e.g. TFT monitors, laptops, LED lighting, etc.). This may well compensate for any further increase in the amount of equipment used. One should therefore consider the possible developments and trends in future uses. Energy concepts have to be sufficiently robust so that they can compensate for the effects of changes. This includes (within certain limits) the flexibility to accommodate changes in use.

Theater floor plan.

Analysis of the heating service in relation to:

a Energy requirements
High demand = dark
Low demand = light

b Internal factors
Favorable conditions = dark
Unfavorable conditions = light

c Resulting energy demand
High energy demand = dark
Low energy demand = light

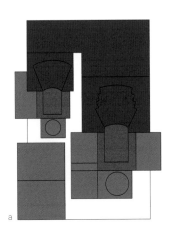
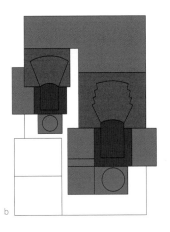
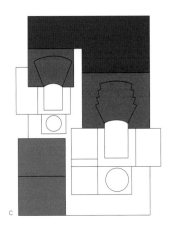

a   b   c

PROPOSAL FOR A METHOD OF ANALYSIS

In order to analyze and optimize zones with respect to energy, an iterative three-step approach can be taken: the requirements of a zone, the factors affecting the zone with respect to position and type, and the resulting energy requirement for maintaining a certain quality are entered separately for the various services. Layout and section drawings with the zones shown are used for the on-going design process.

a) Energy requirements of the use split up according to services, i.e. heating, cooling, lighting, and ventilation. Whether a use has exacting requirements or not should be evaluated. Exacting requirements include small tolerances regarding room temperatures in summer or winter (cooling/heating) or the need for high-quality lighting and ventilation in a room.

b) Energy-related factors for the respective services resulting from the use, the external influences, and the other services. Whether all impacting factors taken together are advantageous or disadvantageous for the provision of services should be evaluated. A high proportion of openings is beneficial for using solar irradiation in winter (requirements for heat and light) but, without additional measures, is disadvantageous for the room temperature in summer (cooling requirement).

c) Expected energy demand for the different services, i.e. heating, cooling, lighting, and ventilation. It should be evaluated whether the computation of requirements and energy-related factors results in high or low energy demands for operating the building.

If the requirements are exacting but impacting factors are favorable, the expected energy demand will still be low. Where impacting factors are not favorable but the requirements are not exacting, it is possible that the expected energy demand will also be low. However, if the requirements are exacting and the impacting factors unfavorable, it is possible that the specified quality of the zone cannot be achieved, or that expensive compensation measures (in terms of services/construction) will be necessary. Where the impacting factors are favorable and the requirements are not exacting, a different use should be considered for this zone, as otherwise the high quality potential of this location could be wasted even though the arrangement itself is not expensive.

The result of this exercise is a number of drawings for the different services, showing areas with exacting requirements or favorable impacting factors. By rearranging rooms according to these potentials it is possible to balance any deficiencies rather than compensating for these deficiencies by technical means. In addition, this process also highlights potential conflicts: a certain area may be favorable for several services (for example, a south-facing position near the building facade with respect to heat and light) but unfavorable in regard to protection from heat in summer (cooling). In that case a zone could be moved, or two areas linked, to make up a deficiency in one from the surfeit in another (e.g. transporting excessive heat to a north-facing zone or one adjacent to an inner courtyard). This is a way of exploiting synergy effects, making independent measures for each individual situation with the resulting high-energy input unnecessary.

PERSON- AND WORK-RELATED REQUIREMENTS

It is primarily the physical comfort of people that determines the requirements of rooms for living, working, teaching, and learning. It is only possible to concentrate and work without making errors, and to recuperate sufficiently, in an environment that is perceived to be comfortable. The well-being of people is based on the perception of a multitude of influences. Owing to individual differences between users and the fact that perception is subjective, it is not possible to achieve one hundred percent user satisfaction. Therefore, optimal impacting factors can only minimize the proportion of those who are not satisfied. The defining parameter is the *predicted mean vote* (PMV), which is a vote by users indicating their evaluation of perceived comfort. This parameter is used to establish the *predicted percentage of dissatisfied* (PPD), i.e. the proportion of those users who are dissatisfied with the internal climate. According to DIN EN ISO 7730, values below 10% are desirable.

USER INFLUENCE AND CONTROLLABILITY

Adjusting the room climate to individual requirements has a significant impact on perceived comfort. Studies show that even the theoretical possibility of individual control increases user satisfaction as users are able to select the climate themselves, within certain limits. There is therefore a psychological dimension in the perception of comfort, which is beyond measurable parameters and is often not considered enough, especially in highly technical concepts. Problems may arise where personal influence is wrongly understood and is seen as an excuse to adhere to disadvantageous habits rather than to question habitual behavior patterns. In this case it is necessary to convey the relevant information in user training sessions or manuals. The architect should see user influence as an opportunity and should replace highly efficient automatic systems, which may however be slow to respond, with individual systems with faster response times. These include windows that can be opened, individual fans, individually controllable solar screening and anti-glare devices, and thermostats and volume flow controls located in each room.

The figure below shows a logical arrangement of factors and perception levels impacting on the sense of comfort. These can be grouped into physical, intermediary, physiological, and psychological conditions.

Most physical conditions can be measured and are covered by construction standards. Ordinances such as the German workplace regulations and legal judgments specify requirements and thereby impact on the design of buildings and services. Often the quality and performance of a building, and the resulting energy requirement, are evaluated on the basis of physical conditions. Intermediary conditions are mostly user-dependent and have to be identified as part of the design process. The nature of activity, the clothing worn, and the number of persons per room have a significant impact on the sense of comfort. Certain habits need to be identified by the architect in conversation with the users; it may be possible to reorganize these habits to achieve a more favorable design. Physiological conditions depend wholly on the individual and are extremely variable. The building must be able to accommodate these flexibly. Exceptions are buildings for specific user groups such as the young, old, or sick, which result in permanent climate requirements at variance to those of the average person.

Finally there are psychological conditions to be considered for the room climate. Surface materials and textures as well as color schemes and the quality of light all impact on the perception of comfort in a room. As a result, rooms may be described as either "cold" or "warm" without referring to the actual measurable thermal qualities. In this way, the interior design can affect the comfort of rooms and the energy required for achieving this comfort through the subjective perception of warm and cold.

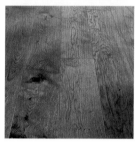

DIN EN ISO 7730 "Ergonomics of the thermal environment: Analytical determination and interpretation of thermal comfort using calculation of the PMV and PPD indices and local thermal comfort criteria".

The model person (nickname: Michel) used by the German weather service also uses the PMV in order to make statements about the well-being of an average person. This "Michel" is 1.75 m tall, weighs 75 kg, has a body surface area of 1.9 m², and is about 35 years old. He is used for determining the perceived external temperature under certain other influences.

1 Comfort levels as they are perceived by people under the influence of different factors.

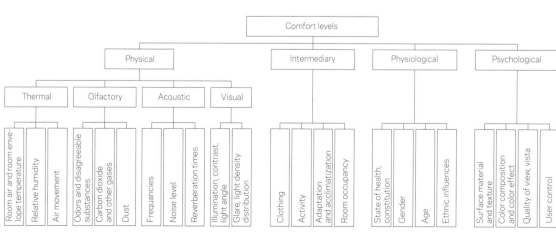

1

2 The sitting person's heat output at a room temperature of 20 °C is just under 120 W. A large part of the heat output is "dry heat", of which 90 W per person is emitted into the room. The heat output increases with higher temperatures, heavier work, or sporting activities; at the same time, the proportion of "humid heat" output increases by evaporation and has to be discharged with the help of ventilation. At temperatures of 28 °C and above, the potential to emit heat is reduced due to insufficient temperature differences between the body and surrounding air. In that case people's well-being suffers.

3 Perceived temperature, depending on the temperature of the room air and the surrounding surfaces. The temperatures of the various building components should be kept within the relevant stated comfort range. This additional restriction limits the ability of building components to contribute to the indoor climate.

Ranges of temperature specified in CEN report CR 1752 in relation to expected levels

Summer (cooling period):
category A 24.5 °C +/- 1.0 °C
category B 24.5 °C +/- 1.5 °C
category C 24.5 °C +/- 2.5 °C

Winter (heating period):
category A 22.0 °C +/- 1.0 °C
category B 22.0 °C +/- 2.0 °C
category C 22.0 °C +/- 3.0 °C

Furthermore, the report includes equations for separately calculating the required external air flow for comfort and health.

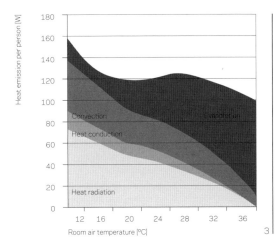

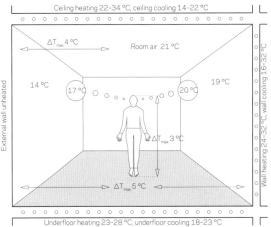

In order to maintain a constant body temperature, people continuously have to give off the heat produced as a by-product of their metabolism. A person's thermal balance, i.e. his interaction with the environment, is based on the following heat exchange mechanisms:
- evaporation through breathing and the skin;
- air convection between the skin and the room;
- thermal radiation to surrounding surfaces;
- thermal conduction to furniture and floor.

These mechanisms can be further subdivided into dry and moist heat output. Only the dry heat output (also called *sensible*) impacts the room directly as a heat source. The moist heat output (embedded in vapor and hence latent) is discharged through ventilation and can only be recovered mechanically. Below is a list of heat output in watts and moisture yield in grams of water per hour for different human activities:
sleeping > 70 W 30 g/h
sitting relaxed > 100 W 30–40 g/h
light work while seated > 125 W 40–60 g/h
light work, standing up > 170 W 60–90 g/h
manual work > 210 W 90–150 g/h
heavy manual work > 250 W 150–200 g/h

The dry component of the heat output can be determined in relation to the room air temperature. The heat input into a room in [Wh/m²d] is calculated from the number of persons and the length of time they are in the room. This is particularly relevant where the occupancy rate is high, and needs to be established, along with the heat input from electrical equipment and solar sources, in order to understand the thermal balance of the room.
When users are content with the temperature, humidity, and air movement in a room, it is considered to have good thermal comfort. The evaluation parameter is the *opera-*tive or perceived temperature. In simplified terms, this refers to the average value of the temperature of the room air and the surrounding surfaces. Depending on the activity, temperatures of 17 °C (for heavy work) to 26 °C (in a bathroom) are perceived as comfortable. Another objective is to minimize the difference between the temperature of the air and the surrounding surfaces. This impacts on the minimum quality of the building envelope and/or the proportion of windows, as well as on the maximum or minimum building component temperatures, if these are used for the transmission of heat. Generally, people perceive heat exchange by radiation as more comfortable than changes in room air temperature. Another point is that the feet should be warmer than the head. This ideal situation can be approximated with underfloor heating in winter and ceiling cooling in summer, while a good compromise throughout the year can be achieved with heating and cooling in the wall.

Target values and maximum fluctuation ranges for operative room temperatures are specified in a number of guidelines. They are different for summer and winter, taking account of external factors, people's acclimatization, and clothing. The CEN CR 1752 standard includes three categories, A, B, and C, which are defined according to user expectations regarding the interior room climate, ranging from high to low. Again, the evaluation parameter is the predicted percentage of dissatisfied (PPD). Building designers must be aware that narrow fluctuation ranges in particular can only be achieved with the help of high-output services installations with fast response times. In order to reduce the amount of services installations, larger deviations can be permitted where window ventilation is possible.

Apart from the perceived temperature, the well-being of people in rooms is determined to a large extent by the quality of the air in these rooms. Adequate fresh air intake, taking into account the quality of the outside air, and the discharge of CO_2, humidity and noxious and odorous substances ensure comfortable conditions.

RELATIVE HUMIDITY

The perception of thermal comfort is closely related to relative humidity. With an interior air temperature of 19 to 22 °C, a relative humidity of 35 to 70% is perceived as comfortable. At higher relative humidity levels there is a risk of condensate, and hence mold formation, at thermal bridges; where levels are lower, there is a risk of increased dust formation, electrostatic charges, and people's mucus membranes drying out. The recommended range for relative humidity is therefore between 40 and 60%.

Irrespective of the relative humidity, it is not possible to create a comfortable environment below a temperature of 16 °C. Above 26 °C, the interior climate is usually perceived as uncomfortably warm, particular in rooms with cooling but without humidity control (active de-humidification of the intake air or surfaces that allow diffusion and can absorb moisture). The so-called *Barackenklima* (German term referring to the interior climate in poorly insulated buildings of lightweight construction) results from excessive temperature and humidity fluctuations in buildings with little thermal storage mass and hygroscopic absorption capability.

AIR MOVEMENT

Air movement also impacts on our sense of well-being. Drafts and turbulence create discomfort while a steady flow of air in summer can help with cooling down the skin and stabilize people's thermal balance. It is important to avoid cold air dropping down near building components with poor thermal insulation or near to overly large window surfaces, and also poorly dimensioned and positioned air inlet apertures.

HYGIENIC AIR CHANGES

People's requirement for fresh air in [m³/h] and their CO_2 output in [l/h] fluctuate depending on their activity and the environmental conditions. The air flow required for providing people with adequate amounts of oxygen is only about one tenth of that required for maintaining the CO_2 concentration at an acceptable level. For this reason, the CO_2 concentration is considered a good indicator for the evaluation of air quality. Typical concentration in undisturbed outside air is 0.04 vol.-%. According to Pettenkofer, controlled ventilation should prevent the indoor concentration from rising by more than 0.10 vol.-% above the outdoor concentration. In accordance with the new DIN EN 13779, the value of 0.14 vol.-% (corresponds to 1,400 ppm) represents a moderate to low room air quality (categories IDA 3 to 4), which however is typical for habitable rooms. In order to achieve this limit of CO_2 content, an average outside air intake of 20 m³/h per person is needed. In the case of intensive activities or where other air polluters such as printers and copiers are present, this value rises to 30–50 m³/h pp. The air exchange should not be increased beyond what is hygienically required (for example, too much intake air) as the increase in air flow requires significantly higher amounts of energy for the air transport and increases the heat loss through ventilation. In spring and autumn it may be necessary to increase the air flow to 40 m³/h in order to discharge humidity, and in winter, the air flow may have to be reduced to 20 m³/h so that the interior air does not become too dry. Further directives regarding air changes are provided by the legally binding *maximum allowable concentration* (MAC), the *workplace exposure limit* (WEL) and the *biological limit values* (BLV).

OLFACTORY COMFORT

Ventilation is also required to discharge odors. For this reason the flow of fresh air required per person, as specified in various standards, includes provisions for discharging unwanted odors. By making appropriate choices for materials, adhesives, and coatings, designers can influence the amount of odor and noxious substances in a room; likewise, users can control air pollution through the selection of electrical equipment (in particular, printers and copiers) and their location. Avoiding sources of emissions can help to minimize the need for air changes and, at the same time, improve user satisfaction. Noxious substances in the room air can be measured in terms of *total volatile organic compounds* (TVOC). Such measurements are conducted by specialist

1   Increasing relative humidity by 10% at normal room air temperature increases the perceived temperature by about 0.3 °C. Where the relative humidity is 60% higher, the internal temperature is therefore perceived to be approx. 2 °C warmer. This effect is further increased with higher internal temperatures. For this reason the comfort zones have been arranged diagonally in the diagram.

In crisis situations, shelters need an air change of 1.8 m³/h per person. This is the minimum required oxygen supply.

DIN EN 13779 "Ventilation of non-residential buildings" with different categories for indoor air quality (IDA).

According to a study conducted by the Fraunhofer Institute, one in five users consider the room climate uncomfortable in naturally ventilated buildings, in partially air conditioned buildings this figure rises to every third person, and in fully air-conditioned buildings, every other person is dissatisfied with the climate.

DIN EN 15251 "Indoor environmental input parameters for design and assessment of energy performance of buildings addressing indoor air quality, thermal environment, lighting and acoustics".

An *olf* is a unit of measure used for indicating the strength of an odor. One olf is the odor emitted by one standard person. Fanger defines such a person as an adult with a skin surface area of 1.8 m² and a hygiene standard of 0.7 baths per day, engaged in a sedentary occupation. The unit of measure not only covers odors from living beings (people, animals, plants), however, but also vapors from building materials.

Standard person	1 olf
12-year-old child	2 olf
Heavy smoker	25 olf
Athlete after activity	30 olf
Interior surfaces (per m²)	
Marble	0.01 olf
PVC/linoleum	0.2 olf
Synthetic fiber carpet	0.4 olf

Perceived air quality is described with the decipol unit of measure. One decipol (dp) is the perceived air quality in a space with a sensory load of one olf, where every second ten liters of fresh air is taken in through the mixed ventilation process.

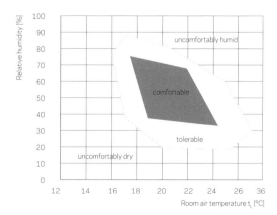

The *illuminance* E is the quotient resulting from the light [lm] arriving at a surface divided by the area of the lit surface. It is measured in lux [lx] or lumen per square meter [lm/m²]. The *luminance* L in candela [cd/m²] defines the light I emitted from such a lit surface of the work area.

The nominal illumination levels to be complied with are specified in DIN 5035, EN 12464 and the German workplace regulations. The illuminance is also included in the use profiles in DIN V 18599 Part 10 as a use-specific requirement, and is used for calculating the energy demand of artificial lighting.

"Absolute glare" is the result of excessive luminance; "relative glare" is caused by excessive contrast between 1) the object seen, 2) the immediate environment, and 3) the distant environment. The maximum contrast ratio of 10:3:1 should not be exceeded.

The daylight quotient D in percent describes the ratio of the internal illuminance (E_i) at a certain point in the layout, at the vertical level of the work place, to the external illuminance (E_a) under an overcast sky.

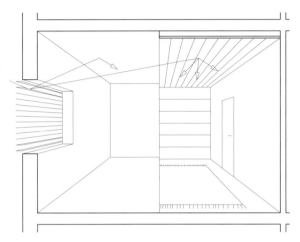

laboratories and should be undertaken shortly after a building is occupied for the first time.

VISUAL COMFORT

People's visual perception accounts for 80 to 90% of their information stimulus. It follows that our perception of the environment depends on our sense of sight. However, people are generally able to adapt to audio and visual environments. While thermal discomfort is not accepted very often, users frequently adjust to unfavorable visual conditions. On the other hand, such conditions clearly lead to a lack of concentration and an increased incidence of errors, and are detrimental to efficient and safe work. For working areas, optimum visual comfort is achieved when the illuminance and the luminance are adjusted to the task in hand. However, the provision of daylight for PC-based work is not without problems. Reflection, glare, and strong contrast lead to a reduction of the visual function and an associated tense sitting posture. For this reason, a solar screening and glare protection device capable of individual adjustment and suitable for the workplace should be available, and workplaces should be sited at a minimum distance from windows. The main viewing direction towards the monitor should be parallel to the window plane. Bright surfaces in rooms reduce the risk of relative glare and, at the same time, increase the utilization of daylight through reflection.

DAYLIGHT AUTONOMY

In our modern everyday world with its artificial lighting we often lack sufficient daylight, which we need to maintain our physical and psychological health. Only daylight has a full spectrum that changes throughout the day and neutral color characteristics. In buildings, the use of daylight must be carefully planned to suit the functions taking place within. The main requirement is for even distri-

bution of light in the room including in areas remote from the windows. The quality of daylight is evaluated using the daylight factor. A workplace with a daylight factor of D > 3% has daylight provision over a period of between 50 and 70% of the day and, with this high daylight autonomy, can make a significant contribution to conserving electricity. Higher daylight factors can lead to overheating of the rooms through solar gain and to glare at the workplace.

ACOUSTIC COMFORT

Acoustic comfort in a room is affected by the amount of protection the room affords from external and internal sources of noise and is best served by a design that suits the functions of the room in terms of room acoustics. The human perception of noise is measured as *sound pressure level* [dB(A)], a psycho-acoustic parameter. This is meant to reflect the combined sensual impression of the physical parameters of sound pressure and frequency. When measuring the *sound reduction index* Rw [dB] of a building component forming part of the building fabric, the frequency dependence of the resistance is replaced by a simplified comparison to a reference curve.

Another measure, the *reverberation time* T [ms], is used to determine the acoustic impression a room makes on the listener. Short reverberation times of 0.3 to 0.8 seconds are beneficial for understanding spoken language, whereas long times of 1.5 to 3.0 seconds improve sonority. DIN 18041 "Acoustic quality in small to medium-sized rooms" makes a distinction between two types of rooms and their speech intelligibility qualities: rooms in which the spoken language is better understood across longer or shorter distances respectively. Depending on the room volume, recommendations are made regarding absorption materials and their positioning in the room. For example, the ceiling of a classroom should have reflecting properties in its center but absorbing properties at front and back. Finally, the assessment of acoustic quality includes properties such as *early decay time* (EDT), articulation class, and others.

COMFORT IN INTERIOR ROOMS

The architect should make a considered choice regarding materials and textures of the surfaces of the room envelope by taking a holistic approach, as one-sided optimizing strategies can lead to serious disadvantages in other comfort categories. This applies to all surfaces in a room, i.e. not only the building components, but also furniture and fittings. For this reason, users should also be involved in the decision-making process in order to ensure that the intended strategy is followed up in the subsequent operation of the building.

REQUIREMENTS RELATED TO GOODS AND PROCESSES

As well as users, work tools, production processes, and even finished products may require special conditions in rooms or may affect rooms in such a way that it becomes difficult to maintain normal requirements. In this case the comfort of users must be weighed against these special requirements and users themselves are called upon to accommodate such situations through choice of clothing and other secondary means.

Conditions resulting from goods or processes that may lead to uncomfortable interior conditions for people working in these rooms, or even constitute a nuisance or risk, include:
- temperature (overheating and excessive cooling)
- humidity (too dry or too humid)
- odors (nuisance)
- noxious substances (risk)
- poisons (risk)
- dust (nuisance, risk)
- pathogens (risk)
- absence of daylight (biorhythm)
- noise (nuisance, risk)
- radiation (risk)
- potential dangers (psychological effect, fear)
- operating times (biorhythm, susceptibility to errors)
- room dimensions (psychological effect, claustrophobia)

Low labor costs, which are the result of optimum performance and high work efficiency (unit labor costs) are an important factor for the commercial growth of companies, particularly in high-wage countries like in Europe. Also, health and safety at work has reached a high level and preventative care is required rather than post-accident care. It is a fact that, in spite of the above, there are workplace conditions that are "contrary to human nature"; this is the responsibility of the operating companies. The conflict here is between making maximum use of investments (e.g. production lines), creating optimum conditions in certain processes (e.g. spray workshops) or

the extreme effects of those production processes (e.g. glass melting), and "normal" working hours as well as the human perception of comfort. Instead of single-shift operation, companies will operate with two or three shifts; instead of providing hygienically adequate room air and comfortable thermal conditions, they will provide operatives with protective clothing and oxygen.

The subjects of heating and cooling and of relative air humidity, and humidity levels in the room should be considered and analyzed in relation to the function of the room. Parameters for determining air changes, as an example, include:
- hygienic air change (provision of oxygen or discharge of harmful substances)
- heating and cooling (input or discharge of heat by ventilation)
- humidity (humidification or dehumidification by ventilation)
- air extraction (discharge of dust and noxious substances, air filtration)

It is not usually necessary to add the air changes required for the different parameters. The most important parameter or the parameter with the highest air change requirement determines the air flow design of the system. The same applies for the heating and cooling parameters. To illustrate the sort of problems a designer may encounter, the following page discusses a number of building and room functions along with their requirements and the resulting conflicts. It is the task of the designer as well as the users to try to minimize factors detrimental to people's comfort.

ELECTRONIC EQUIPMENT STORES

Intensively illuminated showrooms with many electrical items tend to overheat throughout the year. Acceptable conditions can be achieved with a high number of air changes to reflect the number of persons but also to discharge noxious substances outgassing from products. It may enhance sales if products are shown in operation, but at the same time there is a negative effect on the room air which will be detrimental to sales in the showroom. Energy-efficient and low-emission equipment is useful. Greater room heights can be helpful as noxious substances may be discharged with the rising air flow. This effect is further enhanced by extracting the air near the ceiling and introducing fresh air near the floor, for example, through fresh air inlets.

COLD STORAGE AND REFRIGERATED CASES

Rooms for cold storage need to provide uninterrupted cooling for the stored products and this function is only marginally affected by the small heat input from the persons working there and the artificial lighting. The cooling function of these rooms determines the design temperature and relative humidity. Operatives have to adjust to these conditions by wearing appropriate clothing and taking regular breaks. With refrigerated display cases in supermarkets the situation is different. These display cases are opened frequently and are therefore exposed to normal room temperature which, in combination with the intensive lighting, leads to a significant increase in electricity required for cooling. Where the compressors of the cooling aggregates discharge their heat directly into the room air, it is also possible that these refrigerated display cases contribute to overheating.

COMMERCIAL KITCHENS

These kitchens typically have high levels of humidity, odors, and heat input from various appliances, which means that they need constant ventilation. Of necessity, staff are expected to cope with more extreme comfort conditions. In order to avoid contamination of the food with sweat and hair, staff have to wear hats and protective clothing in spite of the high temperatures. It follows that environmental conditions are relatively poor as a result of the requirements for preparation processes, and hygiene regulations allow the staff very little scope to respond by varying their clothing. It is therefore imperative that efficient air conditioning is provided.

SWIMMING POOLS

Compared to standard workplaces, indoor swimming pools are characterized by especially warm and humid conditions. However, the permitted clothing makes sufficient allowance for the situation.

PAINT SPRAYING WORKSHOPS

These rooms are designed as clean rooms so that the quality of the paint coating can be guaranteed and noxious substances can be extracted in a controlled way. The spraying process requires solvent as a liquid carrier for the paint or lacquer. These solvents will evaporate when exposed to heat and have to be extracted through a ventilation system requiring frequent air changes and high-quality filtering. If the process is not an automatic one, operatives have to wear protective clothing (breathing mask, protective overalls) and have to be supplied with fresh air from outside the room. Once the spraying process has been completed, and also sometimes during the spraying process, drying is accelerated by applying additional heat. In this situation it is clear that health and safety at work, and product quality, are considered more important than the perception of comfort. To compensate for the loss of comfort it would be appropriate to provide a staff room and allow for regular breaks.

ELECTRO-MELTING (ALUMINUM, GLASS)

Large power stations, especially nuclear power plants, cannot be started up and shut down at will. In contrast, power consumption in Germany is subject to strong fluctuations during the course of the day. There is only limited scope to make up the difference between demand and supply from rapid response power stations and storage power stations. As a result, the mains network has an excess supply during the night. In order to achieve a better balance by shifting demand from day to night, industrial companies are offered significantly reduced electricity prices during the night as an incentive. In consequence, almost all energy-intensive production of materials from raw materials is carried out during low-demand periods. In this case night shifts are accepted, as the cost of energy dominates the timing of the processes. This situation requires special provisions regarding the organization of work shifts (max. 3 night shifts in a row) and artificial lighting in order to improve safety at work and to "deceive" the biorhythm.

Examples of other processes where conditions are not, or not entirely, determined by people's perception of comfort: mines, airplanes, industrial premises, workshops, joineries, clean rooms for production processes, server rooms, laboratories, operating theaters and laundries.

ANALYSIS OF OPERATING TIMES

The type of use and user group determines the requirements a building has to meet. An analysis of the operating times is carried out in order to establish when and for how long these requirements have to be met. For the design of an energy concept, annual, weekly, and daily operating rhythms and loads are equally important as the requirements themselves. The analysis of the different time periods involved gives the designer some useful indicators and design objectives for the design: for example, the usefulness of storage mass, and hence the choice of materials and surfaces in a room, is determined by the duration of the respective activity over the course of a day, i.e. whether it is ongoing or intermittent.

It is also possible to relate the operating times to the design of services installations, a topic that is also dealt with in the technology chapter. Annual operating patterns of a system have an impact on the selection of certain energy sources; these depend on the external factors, have to provide different levels of service during the different seasons of the year, must be able to provide a base load and cover peaks, and should be utilized on an ongoing basis:

- During the summer period, solar thermal and photovoltaic systems installed on roofs generate just under 80% of the total heat and power they produce. If no relevant demand exists during this period, a consumer must be found or the yield must be stored (the storage requirement ranges from a few hours to seasonal storage). Otherwise this type of system is not suitable.

- Systems using groundwater and the soil can supply heat in winter and cooling in summer. This type of system can be utilized better throughout the year and there is a more balanced use of energy. Of course, a prerequisite for these systems is a demand that approximately matches the energy produced; alternatively, an additional energy source has to be available.

- Not all technologies are able to provide high output at an economic cost. However, this is needed when there is interrupted usage (systems switched off at night, over weekends, and during vacations and holidays). The most common systems capable of producing maximum output are boilers for heat and vapor-compression units for cooling.

- A typical system for meeting base loads is a combined heat and power station. Although these power stations are more expensive to build, they are suitable for meeting ongoing demand on a daily and annual basis.

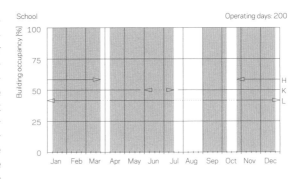

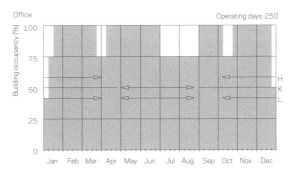

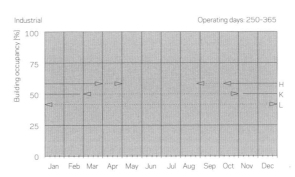

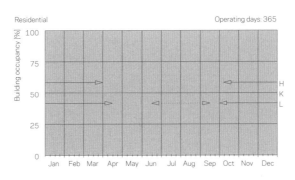

Operational periods have been marked in a dark shade. Depending on the time period depicted, vacation periods, weekday and weekend use, and interruptions and breaks in the course of the day are discernable.

The main heating (H), cooling (K), and ventilation (L) system operating periods over the course of the year are illustrated by arrows for each of the four building uses. Obviously, it is clear that the actual time periods cannot be established until the design concept for the building and its services installations has been completed. However, the operating times analysis provides a suitable basis for an initial analysis of requirements and potentials.

When comparing the annual profiles of operating times of different uses, considerable differentiation becomes apparent. For example, with school buildings it would be worthwhile checking whether the long summer break and the shorter breaks in spring and autumn would adversely affect the viability of solar thermal installations. Certainly, if solar collectors are installed, their orientation should take this into account. Collectors integrated into the facade are advantageous in this situation, as their vertical alignment suits the lower position of the sun. When a photovoltaic system is connected to the mains network it matters less that some yield occurs outside the building's operating times. Nevertheless, it must be taken into account that during the summer break, less of the power produced will be consumed and more will be fed into the mains. Although no energy is wasted, this means the accounting structure changes.

The operating times of a building also provide fundamental input data for programming the services installations. Typical examples are settings for switching off at night and for reduced operation during weekends. It follows that the operating times of a building are not necessarily the same as those of the services. The latter times need to make allowance for heating-up times and possible overruns of usage times.

If the period of analysis is reduced to 24 hours, it is possible to show changing occupancy rates during the day. This allows conclusions to be reached about the specific time periods during which demand must be met. When the daily analysis for different uses is compared, it is possible to assess any potential synergy effects, for example, transferring heat from one zone to another or utilizing storage mass for daytime and nighttime balancing (charge and discharge).

In German schools, almost all classrooms are used from about 8:00 a.m. to 1:00 p.m., and after 2:00 p.m. it is likely that only some of the rooms will be used, whereas in the evening there will be only occasional use of individual rooms by outside agencies or associations. A suitable response to this situation would be to allocate these partial uses to one specific part of the building so that the services can be focused on that area. As can be seen, a significant difference exists between school buildings, with their fragmented use, and office buildings, which are used all day from 7:00 a.m. to 6:00 p.m. and have to fulfill different requirements at different operating times.

In the case of non-residential buildings, there is also the possibility of three-shift operation, which means that nighttime cooling is not an option as the use continues throughout the night. On the other hand, there is no need for a heating-up phase in the morning.

Furthermore, weekly use patterns differ considerably between residential and non-residential buildings. While residential buildings are more intensively used at weekends, the opposite is usually the case with non-residential buildings. Exceptions are shopping centers or sports halls with a six- or seven-day week as their operating pattern.

In these buildings it may be possible to achieve savings through appropriate organization; services must be laid out to meet demand. It is rare that office buildings overheat at the beginning of the week since the building has had a chance to cool down during the weekend, which contributes to comfort levels. Thursday is usually the most critical day of the week.

School
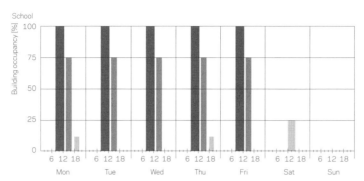
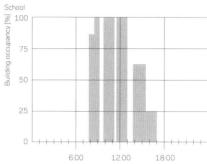

Office
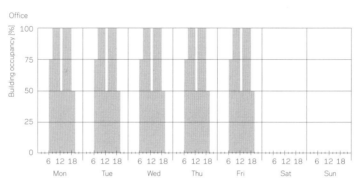
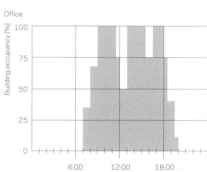

Industrial
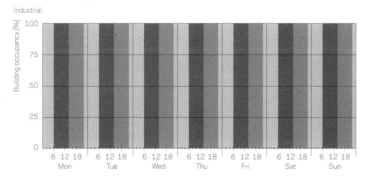
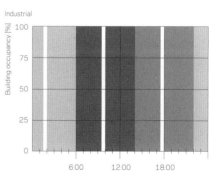

Residential
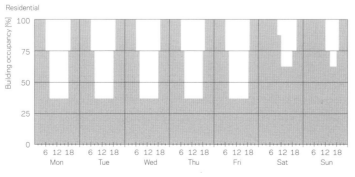
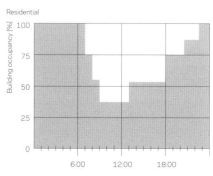

OPERATING PROFILE OF HEATING SYSTEMS

The next two pages show a diagram of an operating time analysis and user-specific characteristics of heating and cooling systems; these are allocated to 42 typical zones of residential and non-residential buildings. The operating profiles show the thermal balance spread over a typical day/typical week in winter (heating) and in summer (cooling).

INTERNAL HEAT SOURCES IN THE COURSE OF A DAY
Internal factors can be derived from the type and intensity of the use. Internal heat results from the presence of persons in the room, the electrical equipment used, and the heat given off by artificial lighting and mechanical ventilation. Depending on the daily use sequence and the specified requirements, the internal sources of heat are distributed by amount and composition, and are illustrated as 24 consecutive columns. In this context the heat emitted by people plays a special role: in contrast to other heat sources, the heat emitted by people does not result in an energy requirement and is independent of technical installations; on the other hand, it varies with the type of activity and number of people present.

HEAT BALANCE IN THE COURSE OF A WEEK
If one considers the heat balance over the period of a week rather than a day, the different columns of the diagram merge together to form a characteristic curve. This *load curve* of the internal factors can then be added to a weekly diagram of the external factors. The wavy lines are a simplified presentation of the specific loads in different buildings, resulting from the heat losses due to transmission and ventilation, and counteracted by solar gains during the day. The analysis during winter is based on the assumption that the weather is dull and cold and fluctuations between day and night are rather small. The heat balance of the building is obtained by superimposing these factors on the use-specific heat input resulting from internal heat sources, assuming that no active system is present. In a situation where the internal heat sources are sufficient to compensate external drops in temperature in winter (approximating the zero line) there is no space heating requirement. The building will be maintained at the required heat level by the heat emissions generated as a result of its use. This situation frequently occurs in very energy efficient buildings with intensive use.

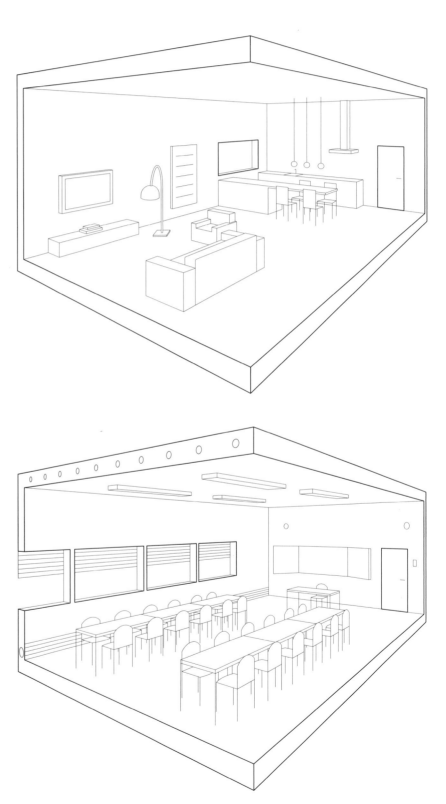

The week diagrams show the possible room conditions occurring in winter:
- in spite of an internal heat output, there is still a requirement for heat to be actively provided by the heating system (light gray area);
- internal heat output and external drops in temperature counteract each other and the room is thermally balanced without active contribution from the heating system (dark gray area touching the zero line);
- the external temperature is low but the internal heat output from use of the room more than compensates for this. This means there is an excess of heat (dark gray area transgresses the zero line).

Intensively used rooms with high occupancy rates, such as offices and classrooms, can produce excess heat from internal sources during winter, provided always that the building has been designed to energy-efficient standards. This means that heat must be discharged in order to avoid overheating during the day. It is preferable that this excess heat is stored at source in building components with appropriate storage mass (this will counteract nighttime cooling); alternatively, it may be actively recovered from the exhaust air and stored or used immediately in another zone.

Even operating periods throughout the day and week have a positive effect, i.e. the heating load is reduced. Depending on the intensity and type of use, there will be more or less demand for heat during winter. The interruption of use during the night and especially at weekends leads to rooms cooling down unless this is counteracted by active measures; a high standard of insulation can minimize heat loss during these periods. This directly affects the design of the heat generation system, heat storage, and means of distribution.

Residential

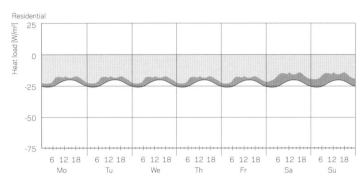

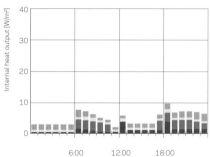

Classroom

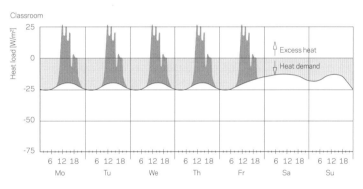

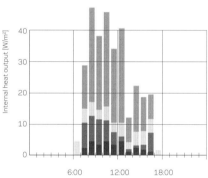

Office

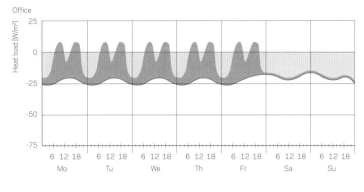

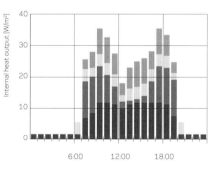

Industrial building

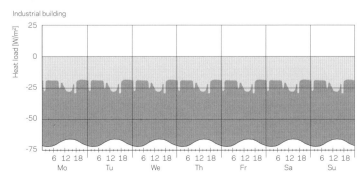

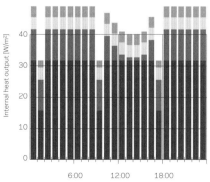

People
Mechanical ventilation
Lighting
Work equipment

OPERATING PROFILE FOR COOLING

In summer, buildings tend to overheat, which means that the maximum target temperature for comfort levels inside is exceeded for prolonged periods. While in winter solar and internal heat gains compensate for heat losses due to transmission and ventilation, these factors contribute to heating up the room during summer. Rooms can only discharge heat to the environment during cool summer nights when there is no solar irradiation and the external temperature drops below the internal temperature.

INTERNAL HEAT SOURCES IN THE COURSE OF A DAY
Heat emissions from people are the same during summer and winter, which is also true for electrical equipment and ventilation. In contrast, the heat emitted from artificial lighting is reduced since there is more daylight during summer and less artificial lighting is needed, provided always that this is not counteracted by any solar screening devices.

HEAT BALANCE IN THE COURSE OF A WEEK
When internal and external factors are superimposed, the problems arising in summer become apparent: rather than counteracting the lack of heat in winter, the effects add up in a non-beneficial way. In the course of the working week, non-residential buildings accumulate heat and heat output is only reduced during times of absence. The analysis is based on a clear summer day with intense solar irradiation during the day and cooling during a clear night. Compared to the cold season, the degree of fluctuation is increased, resulting in a more pronounced day/night rhythm. During the day there may be considerable demand for cooling, while at night the building can benefit from natural cooling. It is important to ensure that heat does not accumulate in the building. This can be achieved through passive cooling at night, as mentioned above, or through active cooling. With active cooling one can also distinguish between daytime and nighttime operation: when cooling is applied directly during the day, users will often complain about uncomfortable conditions (cold drafts, temperature differentials in the room are too high and changing too often). Instead, in buildings with sufficient storage mass, which will heat up only by a few degrees during the day without active cooling, it is possible to shift active and supporting passive cooling to nighttime. The objective is to provide the same thermal conditions every day in the morning and to avoid the building becoming progressively hotter over the course of the week. Where active cooling occurs while users are absent, it is possible to use systems with higher outputs and different transmission elements, which are otherwise considered to provide poor comfort levels (e.g. underfloor cooling).

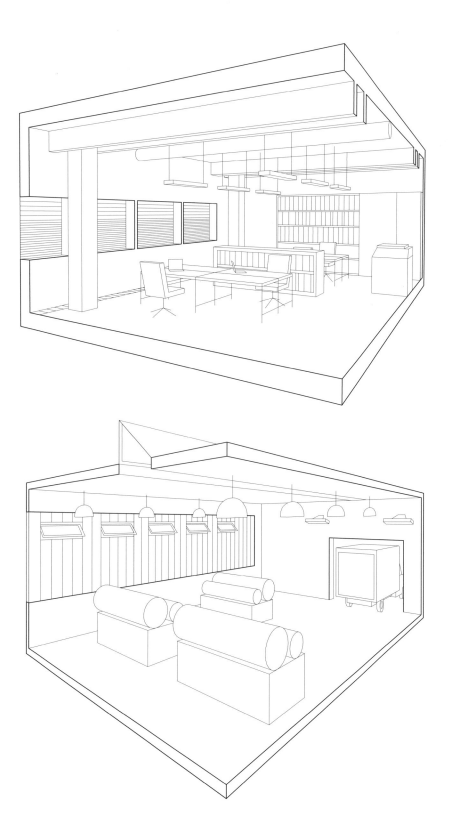

The four axonometric illustrations on this and the previous double page show rooms with different designs and equipment for different occupancy rates and uses. The building envelope, the volume of the room and the window areas together determine the typical load curves for heating and cooling, depending on the occupancy rate, and lighting and equipment densities as internal heat sources.

The operating profile for cooling in summer is also characterized by three different conditions, although with an inverse algebraic sign compared to the analysis in winter:
- despite the internal heat output, the temperature cools down sufficiently during the night, helping to reduce the amount of cooling required during the day—provided there is sufficient available storage mass (light gray area);
- internal heat output and external drops in temperature counteract each other and the room is thermally balanced without active contribution from the cooling system (dark gray area touches the zero line);
- the addition of external and internal heat sources leads to undesirable heating of the room air. If the excess heat is not effectively discharged with the exhaust air, the room is subject to overheating (dark gray area transgresses the zero line).

The presence of excess heat in a room does not necessarily call for active cooling of the building. Particularly in Central European latitudes, residential buildings should be designed such that cooling is not necessary (thermal protection in summer). We must bear in mind however, that the internal temperature increases steadily over the course of the summer period, finally balancing out at a high level. At that stage the room will achieve a balance with its environment over a twenty-four hour cycle: when the internal temperature rises from 22 to 28 °C, the heat emission during the day drops (due to the small difference between the external and internal temperature); at the same time, the potential for nighttime cooling increases.
Where light and dark gray areas are in balance, the maximum internal temperature has been reached.

- ▇ People
- ▨ Mechanical ventilation
- ▨ Lighting
- ▇ Work equipment

Residential

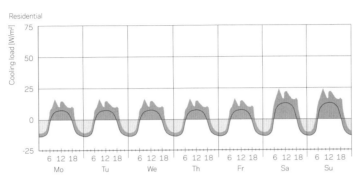

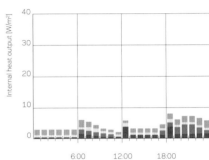

Classroom

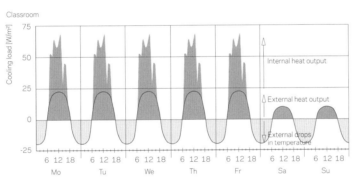

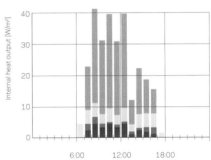

Office

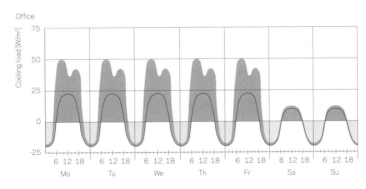

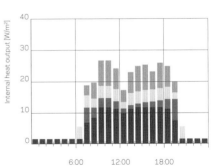

Industrial building

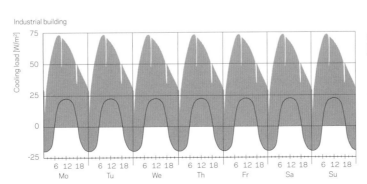

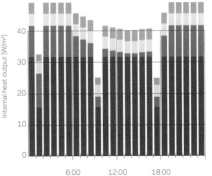

ENERGY PROFILE OF THE DIFFERENT ZONES

The diagram represents the internal factors for 42 uses and ancillary areas in residential and non-residential buildings. This is based on use profiles in accordance with DIN V 18599 Part 10 (2007) and Part 100 (2010), which are summarized and revised, and extended by parameters and the author's calculation results.

Actual figures, further diagrams and tables for the energy profiles of the different use zones are included in the appendix. The tables show both fixed values as specified in the relevant standards, as well as parameters that depend on different types of buildings and design concepts. With the help of the specified space requirement per person, it is also possible to estimate the main use areas to be created, where only the number of users is known.

CREATING ZONE PROFILES

The tables contain structural parameters, energy parameters, and derived reference values. All data are based on assumptions which have to be checked and confirmed, or adjusted. Most data have an upper and lower limit value in order to indicate a typical range. Where future properties have not yet been fully determined, it is probably most useful to use the average value. Zone profiles can be used for initial, approximate calculations as well as for developing a strategy. For example, in situations with long hours of use during daytime and high internal heat emissions, special attention must be paid to the design of solar screening and control of daylight. Where the use extends into the night, the efficiency of the artificial lighting becomes more important. At the same time, the potential for using storage mass in combination with natural cross ventilation is reduced. If rooms are frequently unoccupied for short periods during the day, it may be appropriate to control the demand by presence sensors. The same applies to the use of building control technology for optimizing operating conditions during medium-term and long-term interruptions of use.

ZONE PROFILE CHARACTERISTICS

The operating times and days result in values for the operating hours of the building during day and night. A distinction is made between five-day, six-day, and seven-day weeks, and operating breaks are taken into consideration (school holidays, unoccupied periods in lecture halls). These operating times also apply to the services installations, albeit with lead times of between 30 minutes and 2 hours per day, in line with standard guidelines or the design concept. It is also important to note the night-time requirements for buildings that are used over a 24-hour period (residential) and the fact that some buildings with intermittent use (museums) may need full-time operation.

Detailed presence data give information about the hours and frequency a user is actually present in the allocated room. The values are lower for smaller rooms allocated to individual users and special rooms, compared to larger rooms. Presence control mechanisms can be used to reduce the requirement for artificial lighting and mechanical ventilation when users are absent. In addition it should be noted that there is no heat emission from users and user-operated equipment during these periods. The space requirement per person depends on the type of use and building. There are certain guide values which should be selected to suit the respective design project. The same applies to the clear height of rooms and the resulting air volume per person.

USE-RELATED HEAT INPUT

The heat output rate defines the maximum heat output per square meter in mechanically ventilated, artificially lit rooms occupied by people and their work equipment. For electrical equipment, the heat output rate results from the power needed for its operation less a proportion not emitted into the room. In addition, the heat balance of the room is affected by heat emitted from the bodies of users. The diagram shows how the heat output varies in rooms depending on the type of output and the intensity of their use. The resulting heat yield per day is calculated by multiplying the heat output rate by the daily hours of use/operation. This includes periods when people are absent and no equipment is used. The operating profiles illustrate the changing heat input over the course of days, weeks, and years.

APPLYING ZONE PROFILES

Where the functional and space requirements for a building have been provided, it is possible to calculate the energy requirements of the different uses by multiplying the respective areas by the rates per unit area for that use. This makes it possible to identify zones that can be used to heat other areas because of the excess heat produced in them, rather than simply discharging that heat to the outside through the ventilation system. The appendix to this book includes further tables and diagrams relating to zones. The net and primary energy rates listed in these tables and diagrams can be used to produce an approximate estimate of the typical resulting energy requirements for the respective building standard.

People and equipment as well as the necessary artificial lighting and ventilation in rooms all produce heat, which is given per square meter of usable area for calculation purposes.

The following factors were taken into consideration for the calculation of the heat produced by different sources:

Persons = use period × occupancy rate × heat output depending on degree of activity

Work equipment = use period × power input of equipment

Lighting = use period × power input of light fittings less daylight input (solar input is not shown as part of daylight)

Ventilation = operating period × power input of intake fans (drops or rises in temperature due to temperature differentials between internal and external temperature are not shown)

The use period has been divided into daytime and nighttime hours. The distribution is relevant for the degree of daylight provision.

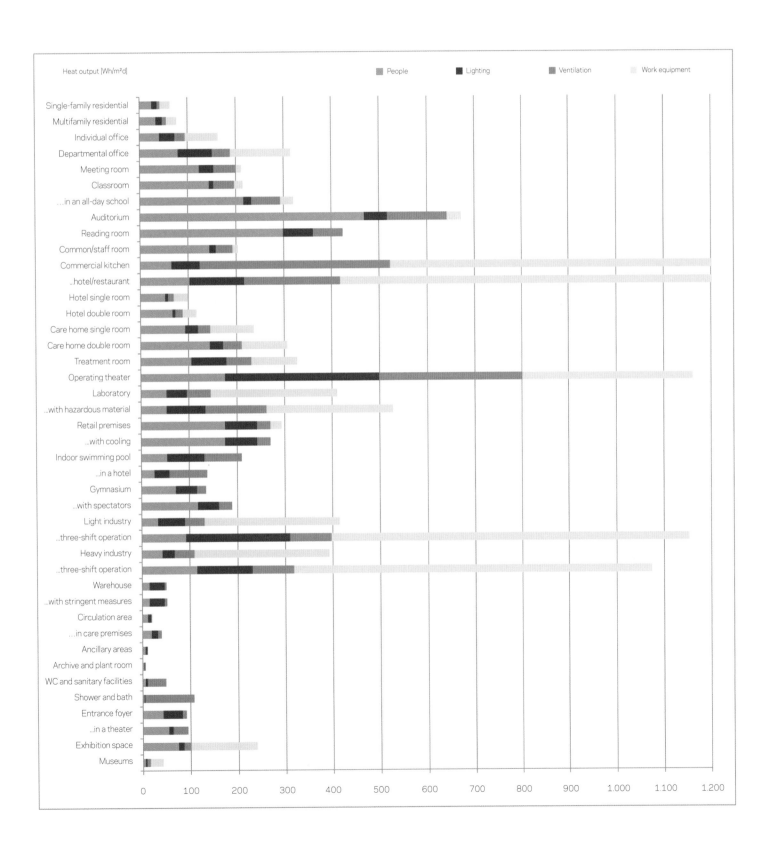

Heat output [Wh/m²d]

Legend: People · Lighting · Ventilation · Work equipment

Single-family residential
Multifamily residential
Individual office
Departmental office
Meeting room
Classroom
…in an all-day school
Auditorium
Reading room
Common/staff room
Commercial kitchen
…hotel/restaurant
Hotel single room
Hotel double room
Care home single room
Care home double room
Treatment room
Operating theater
Laboratory
…with hazardous material
Retail premises
…with cooling
Indoor swimming pool
…in a hotel
Gymnasium
…with spectators
Light industry
…three-shift operation
Heavy industry
…three-shift operation
Warehouse
…with stringent measures
Circulation area
…in care premises
Ancillary areas
Archive and plant room
WC and sanitary facilities
Shower and bath
Entrance foyer
…in a theater
Exhibition space
Museums

0   100   200   300   400   500   600   700   800   900   1.000   1.100   1.200

USE-SPECIFIC INTERNAL HEAT SOURCES

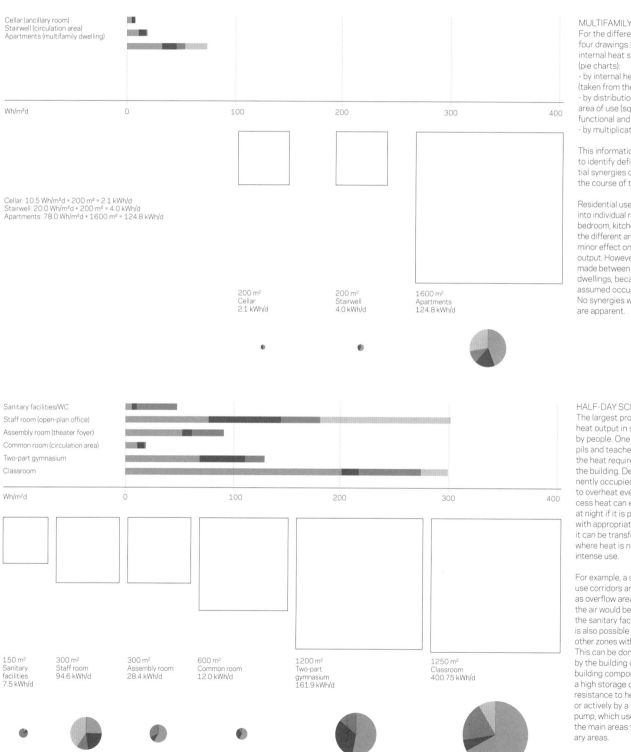

Cellar (ancillary room)
Stairwell (circulation area)
Apartments (multifamily dwelling)

Wh/m²d   0   100   200   300   400

Cellar: 10.5 Wh/m²d × 200 m² = 2.1 kWh/d
Stairwell: 20.0 Wh/m²d × 200 m² = 4.0 kWh/d
Apartments: 78.0 Wh/m²d × 1600 m² = 124.8 kWh/d

200 m²
Cellar
2.1 kWh/d

200 m²
Stairwell
4.0 kWh/d

1600 m²
Apartments
124.8 kWh/d

MULTIFAMILY DWELLING

For the different building uses, the four drawings show how absolute internal heat sources were created (pie charts):
- by internal heat sources per area (taken from the energy profiles);
- by distribution of absolute square area of use (squares depicting the functional and space requirements);
- by multiplication.

This information can then be used to identify deficiencies and potential synergies of the zones during the course of the year.

Residential use is not subdivided into individual rooms (living, dining, bedroom, kitchen, bathroom, etc.) as the different areas only have minor effect on the resulting heat output. However, a distinction is made between single and multifamily dwellings, because this affects the assumed occupancy rate.
No synergies with other areas are apparent.

Sanitary facilities/WC
Staff room (open-plan office)
Assembly room (theater foyer)
Common room (circulation area)
Two-part gymnasium
Classroom

Wh/m²d   0   100   200   300   400

150 m²
Sanitary facilities
7.5 kWh/d

300 m²
Staff room
94.6 kWh/d

300 m²
Assembly room
28.4 kWh/d

600 m²
Common room
12.0 kWh/d

1200 m²
Two-part gymnasium
161.9 kWh/d

1250 m²
Classroom
400.75 kWh/d

HALF-DAY SCHOOL

The largest proportion of internal heat output in schools is generated by people. One might say that pupils and teachers together produce the heat required for heating the building. Densely and permanently occupied rooms may tend to overheat even in winter. The excess heat can either be utilized at night if it is possible to store it with appropriate storage mass, or it can be transferred to other zones where heat is needed due to less intense use.

For example, a solution may be to use corridors and assembly rooms as overflow areas for exhaust air; the air would be extracted through the sanitary facilities. However, it is also possible to transmit heat to other zones without air as a medium. This can be done either passively by the building components (solid building components with a high storage capacity and low resistance to heat conduction), or actively by a discharge air heat pump, which uses excess heat from the main areas to heat secondary areas.

ADMINISTRATION BUILDINGS

The daily heat output from internal heat sources in open-plan offices measuring 1,800m² in size almost equals the quantity of heat produced by burning 60 liters of heating oil. While this substantially reduces the energy requirement for heating in winter, the heat output inevitably leads to overheating in summer. If the internal temperature of the offices is controlled, adjacent circulation areas do not require any heating/cooling. In spite of the fact that these areas take up a large proportion of the building, their impact on the thermal balance is low because not much heat is emitted from internal sources.

Meeting rooms present a special case. They are heavily affected by heat from people, with significant fluctuations throughout the course of the day. The same applies to air changes, whether these are controlled according to the occupancy rate or measured CO_2 values. In this zone it makes sense to use the intake air also for thermal purposes. Requirements of offices are more often determined by the heat emitted from electrical equipment. Where cooling from intake air is not sufficient, it is necessary to make use of building components.

INDUSTRIAL BUILDINGS

The thermal balance of industrial buildings is determined by the production process. This is true for the heat output from machines and equipment, and also for the heat loss due to ventilation. In heavy industry, ventilation is usually done by natural means, since the amount of electricity and maintenance needed for mechanical systems would be excessive. The emphasis is on the discharge of dusts and noxious substances; heat generated in the processes largely compensates for heat loss.

In light industry, the thermal balance is not much different, but lighting and ventilation may need closer attention. Where ventilation is supported with mechanical means, the heat content of the discharge air can be extracted, e.g. with the help of exhaust air heat pumps, and used for heating adjoining areas or for preparing domestic hot water.

IMPACTING FACTORS AND ENERGY CONCEPTS FOR DIFFERENT BUILDING TYPES

Residential buildings shelter people. Occupants range from retirees and working singles to families with children, and the user-specific requirements vary accordingly. Furthermore, concepts must remain flexible to accommodate future changes in use. An average of 35 m² residential space is currently available per person in Germany. This space comprises rooms for living, working, sleeping, eating, and cooking, as well as sanitary facilities. The design temperature for living rooms is 19-21 °C, for bedrooms 16-18 °C, and for bathrooms 24-26 °C for short periods.

USE ZONES
A single zone model with normally heated rooms os used for balancing, since the impacting factors and comfort requirements are uniform. Nonetheless, rooms should be oriented toward the sun as appropriate. ⌐ **chapter 4** A distinction can be made between use intensities and occupancy rates of single or multifamily dwellings.

HEATING
The heating requirement depends greatly on the building's age. Until the 1970s, approx. 300 kWh/m²a was required. In 1995, the German Heat Insulation Ordinance (WSchV1995) stipulated a value of 54-100 kWh/m²a for new buildings. Since 2009, the Energy Conservation Regulations (EnEV 2009) stipulate 30-60 kWh/m²a for space heating, depending on the building type.

DOMESTIC HOT WATER
The typical domestic hot water demand in a multi-person household is 30 l per person per day. Taking a standard temperature of 45 °C, the resulting net heat requirement is 400 kWh/a per person, or approx. 12.5 kWh/m²a at average occupancy rates. Due to losses incurred in generation, storage, and distribution, the final demand is often twice as much.

COOLING
Residential buildings should be designed in a way that obviates active cooling; this is verified by calculating the amount of thermal protection in summer. Impacting factors are orientation-dependant solar irradiation and any screening devices.

CONTROL AND USER INFLUENCE
Typical control mechanisms include thermostatic radiator valves, room thermostats, and external temperature sensors for controlling the heat generation. Automation is increasingly used for new construction, yet it is hardly found in existing buildings. There are no set times for which comfortable conditions must be provided.

NATURAL AND MECHANICAL VENTILATION
Typical residential uses require an average air flow of 18 m³/h per person to maintain an acceptable CO_2 concentration level. Depending on the degree of activity, this value varies between 12 m³/h per person (bedrooms) and 22 m³/h per person (utility rooms). For the intermittent discharge of noxious substances, odors, and humidity, this value is increased to 20-30 m³/h per person. Mechanical ventilation systems can be used to recover waste heat, discharge humidity, and ventilate rooms in a controlled and constant way.

NATURAL AND ARTIFICIAL LIGHTING
At daytime, natural lighting should suffice. Internal rooms and areas without access to daylight should be minimized; potential exceptions are plant rooms, storage, and WCs. Energy-efficient luminaires should be used if not in conflict with considerations such as light temperature or frequency of switching.

ELECTRICAL EQUIPMENT
Typical equipment includes household and small appliances, consumer electronics, artificial lighting, and building services. In optimized buildings, the primary energy consumed directly by users now exceeds what is needed for heating, ventilating, and lighting. Nevertheless, this aspect is not covered by the EnEV provisions.

EFFECT OF BUILDING TYPES
The typology varies from detached houses to row houses and apartment blocks, etc. The energy required for operating the building depends on its compactness, construction method, and the orientation and characteristics of the building envelope. Greater compactness and a southern exposure improve the energy balance

COMMON RESIDENTIAL ENERGY CONCEPTS
With residential buildings, providing for space heating is still of primary concern and lawmakers are focused on its reduction and efficient provision. In the early 1990s, the Passivhaus concept reduced heat loss to a level at which space heating could be achieved solely by heating the physiologically required intake air. Other contemporary concepts include Net Energy, Zero Energy, and Plus Energy houses, which aim at a balanced annual energy demand or even provide excess energy to be fed into the supply networks. Subsidies tied to the stipulations contained in the current EnEV are available from federal and state governments (e.g. KfW Effizienzhaus).

EnEV standard for new construction:
Heating load 30-60 W/m²
Avoid the need for cooling
Heat for space heating
30-60 kWh/m²a
Ambient cooling to be avoided
Domestic hot water 12.5 kWh/m²a
Primary energy 40-100 kWh/m²a
(without lighting)

Passivhaus standard:
Heating load 9-10 W/m²
Avoid the need for cooling
Heat for space heating
12-15 kWh/m²a
Ambient cooling to be avoided
Domestic hot water 12.5 kWh/m²a
Primary energy max. 40 kWh/m²a
(incl. electricity 120 kWh/m²a)

EnEV standard for new construction:
Heating load 25–50 W/m²
Cooling load 25–100 W/m²
Heat for space heating
20–50 kWh/m²a
Ambient cooling 20–80 kWh/m²a
Domestic hot water 2–5 kWh/m²a
Primary energy 100–300 kWh/m²a
(incl. ventilation and lighting)

Passivhaus standard:
Heating load 5–10 W/m²
Cooling load 5–15 W/m²
Heat for space heating
5–15 kWh/m²a
Ambient cooling 10–20 kWh/m²a
Domestic hot water 2–5 kWh/m²a
Primary energy max. 40 kWh/m²a
(incl. electrical equipment
120 kWh/m²a)

Office buildings are workplaces where people perform so-called light activities in a seated position. The spectrum of users and the use periods are fairly clearly defined. The average floor space per workplace is 15 m²; there is a tendency for this to increase slightly, in spite of flexible workplace concepts. Office equipment is subject to similar requirements and also technical developments which may involve changes.

USE ZONES
Usually the energy balance is based on a multi-zone model, as different use intensities and requirements in the different areas frequently lead to different services installations. The determining main zones are cellular (individual) and departmental offices. If these take up more than two thirds of the space, it is also possible to use a single zone model based on a departmental office (open plan). Seminar and meeting rooms occupy less floor space but require more sophisticated ventilation and cooling, depending on the occupancy level. Other uses are circulation areas, which may be extended into combination zones, kitchenettes, and sanitary facilities. Functions such as canteens, kitchens, and server rooms need to be considered separately.

HEATING
In existing buildings the provision of heat for space heating constitutes a major part of the energy requirement. Modern office buildings with appropriate thermal insulation and heat recovery are largely able to cover their heat requirement from internal heat sources (people) and office equipment. Compared to residential buildings, it is therefore easier to achieve the Passivhaus standard.

DOMESTIC HOT WATER
The need for a domestic hot water supply should be ascertained. In view of the small demand and long pipelines, centralized systems lead to high distribution losses and can be problematic with respect to Legionella contamination. Therefore in kitchenettes, decentralized electric boiler systems should be considered. Unless required for comfort, hand wash basins should not be supplied with hot water. Hot water should however be provided where showers are required (e.g. for cyclists, particularly within inner city areas).

COOLING
Whereas the trend in heating is to reduce demand, the demand for cooling is on the increase. Effective, usually movable solar screening, which allows the provision of daylight at least in the upper third of the window, and efficient lighting and office equipment reduces internal heat emissions and hence the demand for cooling.

CONTROL AND USER INFLUENCE
Where there are strong fluctuations, comfort levels can be better achieved when users are given the option of individual control. This may lead to savings in services installations but can also create conflict and waste if misused. Users generally like to control the opening of windows, solar screening and anti-glare devices, and individual workplace lighting, as well as heating/cooling.

NATURAL AND MECHANICAL VENTILATION
In order to minimize the energy required for mechanical ventilation, it is worth considering natural ventilation in offices even though this may lead to increased heat losses or gains. The zoning and facade design should allow for natural ventilation, at least during some of the seasons. It also makes sense to use cross ventilation, which benefits from the pressure differential between different sides of a building. For this purpose it is necessary to provide overflow areas, connect atria and shafts, and provide options for night ventilation, including intruder-proof air inlets. Natural ventilation can be supported with air extraction systems, where heat recovery can be included using exhaust air heat pumps.

NATURAL AND ARTIFICIAL LIGHTING
Office work generally takes place during the day, which means that the potential for the utilization of daylight is high. Regarding artificial lighting, the primary energy balance only covers those light fittings which are permanently installed. However, beyond the primary energy balance, a flexible lighting system may lead to savings in electricity and, secondly, reduce the cooling load.

ELECTRICAL EQUIPMENT
The heat emitted from office equipment has to be entered on the demand side of the energy balance.

EFFECT OF BUILDING TYPES
A compact building design reduces the area of the building envelope and hence investment costs, but it also reduces the potential for using natural lighting and ventilation.

COMMON OFFICE ENERGY CONCEPTS
Building component activation for heating/cooling, supported by fast-acting control mechanisms (conditioned intake air, radiators). Due to the amount of internal heat emissions, the requirement for space heating is reduced. The energy balance of offices is affected more by items such as daylight control, solar screening and unobstructed view, the choice of interior surfaces in relation to storage capacity and reflectivity, intelligent ventilation concepts, and the efficiency of office equipment.

Schools are primarily for learning; however, as a result of the trend towards all-day schools in Germany, other room qualities are becoming increasingly important (staff and common rooms, catering facilities). The degree of activity and level of equipment largely depend on the actual use of the room. While classrooms typically have high occupancy rates while in use, there are also long periods throughout the day, week, and year (holidays) when they are not in use.

USE ZONES

Classrooms make up the main zone, and specialized classrooms may constitute separate areas depending on their equipment and use (chemistry, physics). Circulation areas are often enlarged to provide space for breaks and "hanging out," and there are also assembly halls and administration facilities. Gymnasiums are usually separate buildings.

HEATING

In existing buildings, supplying space heating constitutes a major part of the energy requirement. Modern school buildings with appropriate thermal insulation and heat recovery are largely able to cover their heat requirement from internal heat sources (people) and solar gains. In these buildings, the function of heating systems is confined to preheating the rooms when they have been unoccupied for a while, rather than to continuous heating.

DOMESTIC HOT WATER

As in office buildings, the provision of domestic hot water (DHW) should be reduced to the necessary minimum, with gymnasiums being the exception. A central DHW system involves substantial losses in distribution and storage of the hot water, so decentralized DHW preparation should be considered. If the gymnasium is also used in summer, it is worth examining the option of solar heat.

COOLING

Cooling is not so much of a problem in German schools as the use periods are from morning to early afternoon, and the buildings stay closed in the summer during the hottest period; in addition, pupils can be sent home when the heat rises to an excessive level. To avoid overheating during summer, the building should have sufficient storage mass (> 100 Wh/m²K) in its rooms (avoid suspended ceilings, solid concrete construction) and have a means of natural cooling (nighttime ventilation). It is also important to consider the room acoustics; one option is to install back-ventilated sound absorbing materials at ceiling perimeters. Rooms with high internal heat output (IT classrooms) should be located at the north-facing facade or in basements.

CONTROL AND USER INFLUENCE

Central building control technology is appropriate in view of the intermittent use patterns and the allocation of rooms to specific uses. Room thermostats with the option of manual fine temperature adjustments of +/- 2 °C are beneficial. Windows should have opening casements and be fitted with sensors for the ventilation and heating systems.

NATURAL AND MECHANICAL VENTILATION

In view of classroom occupancy rates, the required air quality cannot be attained without mechanical ventilation. Even where ventilation systems are used, natural ventilation should be part of the concept. In classrooms, a minimum of 0.1 m² of opening casement should be provided per seat where cross ventilation exists, and a minimum of 0.3 m² where there is no cross ventilation.

NATURAL AND ARTIFICIAL LIGHTING

Important for the energy balance are the utilization of daylight and the efficiency of artificial lighting, as well as the construction and quality of surfaces. The minimum reflectance of interior surfaces should be 0.80 for ceilings, 0.50 for walls, and 0.30 for floors, unless other use requirements call for different reflection rates.

ELECTRICAL EQUIPMENT

Classroom equipment varies, but computer use is steadily increasing. Energy efficient units (laptops) use less energy and produce less unwanted heat.

EFFECT OF BUILDING TYPES

There are a number of quite different building types, ranging from compact building cubes to very spread out buildings. It is important that the energy concept is adapted to the structural, internal, and external factors.

COMMON SCHOOL ENERGY CONCEPTS

The application of the Passivhaus concept is appropriate. This requires that nighttime cooling is minimized through high levels of thermal insulation, a fairly airtight building envelope, and a highly efficient ventilation system with heat recovery. Once lessons have started, the heat emitted by users is sufficient to cover the heating requirement; any excess heat is discharged with the exhaust air. Since the air flow is relatively high to cover the physiological requirements, the air quality is primarily controlled through the ventilation system. It is also possible to make use of building components in order to stabilize the interior room climate. Owing to the relatively high cost of installing and running an active cooling system, such systems are not often specified. The Passivhaus construction method can also be used for gymnasiums.

EnEV standard for new construction:
Heating load 30–60 W/m²
Cooling load 20–50 W/m²
Heat for space heating
30–50 kWh/m²a
Ambient cooling 20–40 kWh/m²a
Domestic hot water
10–20 kWh/m²a
Primary energy 75–250 kWh/m²a
(incl. ventilation and lighting)

Passivhaus standard:
Heating load 5–10 W/m²
Cooling load 5–10 W/m²
Heat for space heating
5–15 kWh/m²a
Ambient cooling 5–15 kWh/m²a
Domestic hot water 5–15 kWh/m²a
Primary energy max. 40 kWh/m²a
(incl. electrical equipment
120 kWh/m²a)

EnEV standard for new construction:
Heating load 25-100 W/m²
Cooling load 0-100 W/m²
Heat for space heating
25-150 kWh/m²a
Ambient cooling 0-100 kWh/m²a
Domestic hot water 5-25 kWh/m²a
Primary energy 100-500 kWh/m²a
(incl. ventilation and lighting)

Passivhaus standard:
(not fully defined)
Heating load max. 10 W/m²
Cooling load max. 15 W/m²
Heat for space heating max.
15 kWh/m²a
Ambient cooling max. 15 kWh/m²a
Domestic hot water 5-25 kWh/m²a
Primary energy max. 40 kWh/m²a
(energy efficient equipment)

There is an extensive range of different factories and production buildings. Depending on the type of use and method of production, some buildings may have fairly high occupancy rates, whereas others, with highly automated production processes, may only need a few operatives in attendance. Another aspect is the type and intensity of activity carried out in these buildings. These are defined in degrees of activity, which in addition to the requirements regarding temperature and relative humidity, determine the internal heat output generated by people. The use periods of these buildings depend largely on the shift patterns, with common patterns comprising either one, two or three shifts.

USE ZONES
Owing to the wide range of different process and product requirements, there is also a wide range of use zones with corresponding impacting factors. As a primary consideration, industrial buildings can be grouped into warehouses and production buildings. Beyond that it becomes apparent that the multitude of functional characteristics makes it impossible to define different types of commercial activities in the form of standardized use profiles. However, certain basic factors can be identified which help in defining energy zones and making assessments, and which also make it possible to derive appropriate use profiles.

HEATING
When people are engaged in heavy physical work, lower room temperatures are still perceived as comfortable. However, even these lower limits may be difficult to maintain, for example where finished goods are continually moved to the outside, or raw materials are brought into the building from the outside. One such example is that of a train driven into a repair workshop for repair or maintenance at below-zero temperatures; this exposes operatives to severe working conditions at very low temperatures. Where it is difficult to ensure appropriate comfort levels indoors, it is possible to provide localized heating. Another method of responding to these factors is to wear appropriate clothing.

DOMESTIC HOT WATER
Industrial premises often have showers as part of their staff sanitary facilities which need DHW throughout the year. The quantity depends on the type of work (type of activity and associated soiling) and shift operation.

COOLING
Active cooling is only used in buildings with special goods and processes and almost never for reasons of comfort. The reason is that the amount of heat generated internally from large equipment and production machinery (for example, glass melting) is simply too great.

CONTROL AND USER INFLUENCE
Individual control mechanisms are not normally provided in commercial buildings.

NATURAL AND MECHANICAL VENTILATION
It is important to ascertain whether the storage or processing of materials in commercial or industrial premises involves the release of noxious substances. It may be necessary to install a mechanical ventilation system in a building where it is possible that the *maximum allowable concentration* (MAC) of noxious substances is exceeded, or for other technical reasons. In all other cases industrial buildings are normally ventilated naturally.

NATURAL AND ARTIFICIAL LIGHTING
There is also a wide range of lighting requirements. Natural lighting can be admitted on two sides through the facade or skylights (preferably north-facing).

ELECTRICAL EQUIPMENT
The amount of heat given off in such buildings varies substantially depending on the type, size, and number of plant and machinery. Any heat emitted by such equipment has to be dealt with by the ventilation system. Very large equipment may have cooling circuits, the waste heat from which can be used to cover the heat requirement for space heating and DHW preparation.

EFFECT OF BUILDING TYPES
The height of a building is an important factor for its lighting and ventilation. Furthermore, the type of air handling system depends to an extent on the geometry of the building. The height of these buildings can range from 4 m (e.g., small production workshops such as joineries, car repair workshops) to 18 m (high bay warehouses, large industrial production facilities).

COMMON INDUSTRIAL ENERGY CONCEPTS
Depending on the internal factors in these buildings, a number of different energy concepts have been developed. The main objective is to benefit from synergies. This includes the use of waste heat from production processes for heating adjoining areas.

HEAT I COOL
BUILDING

COMPACTNESS AND ORIENTATION

COMPACTNESS

The compactness (the ratio of the surface area to the volume) of a building has an influence on the energy requirements of the building. ↘p.32 However, the extent of this influence depends on the geographical location and the use of the building. In climates with generally high ambient temperatures and no cold winters (e.g. desert climates) or for buildings with high internal heat loads, the degree of compactness is of secondary significance. Reduced compactness can then even favor the natural ventilation and cooling of the building. In Central Europe however, a high degree of compactness is useful. From a design viewpoint, achieving a compact building means avoiding indentations (e.g. recessed balconies), projections (e.g. bay windows), sharp angles in the heat-conducting building envelope, and simplifying the shape of facades and roofs. ↘1 In the case of a sectionalized design, indentations, projections, or atriums need to be used in a conceptually and structurally meaningful way to compensate for the increased losses, e.g. by means of higher efficiency insulation or a covered courtyard (unheated buffer area). As a rule, larger buildings are more compact and therefore lose less heat through the building envelope. ↘2

ORIENTATION

The orientation of a building determines the possibilities for making use of solar energy, either by reducing the central heating requirement (passive solar energy), or directly by means of photovoltaic systems or solar thermal energy (active solar energy). The sum of the annual useful solar radiation depends on the orientation of the facade surfaces and the location. ↘3 However, the daily amount of radiation striking the building fluctuates greatly because of the changing angle of the sun throughout the seasons. For this reason, openings and solar technology need to be arranged so as to maximize solar yield for the building's operational requirements. The compactness, and the consequent thermal losses from the building shell, the percentage of openings in each facade related to the orientation of the building, as well as the latitude and climate, are the criteria that determine the shape of the building and its orientation.

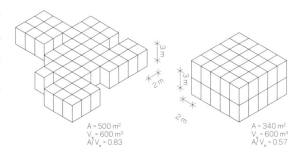

A = 500 m², V$_e$ = 600 m³, A/V$_e$ = 0.83

A = 340 m², V$_e$ = 600 m³, A/V$_e$ = 0.57

1 Both buildings: gross floor area = 200 m², usable floor area approx. 160 m²

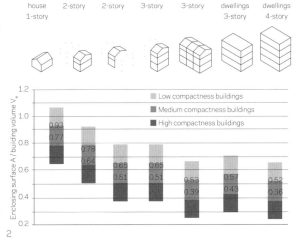

| Single family house 1-story | End-row house 2-story | Mid-row house 2-story | End-row house 3-story | Mid-row house 3-story | House with multiple dwellings 3-story | House with multiple dwellings 4-story |

Enclosing surface A / building volume V$_e$

- ▨ Low compactness buildings
- ▨ Medium compactness buildings
- ▪ High compactness buildings

0.93 / 0.77 | 0.78 / 0.64 | 0.65 / 0.51 | 0.65 / 0.51 | 0.53 / 0.39 | 0.57 / 0.43 | 0.52 / 0.36

2

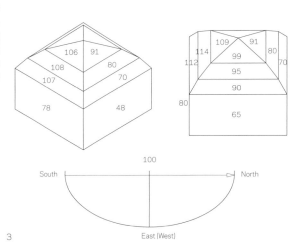

South — 100 — North

East (West)

3

1 Varying A/V$_e$ ratios depending on the layout of the building. In a building with an elongated layout the area of the heat-conducting building envelope is 48% greater than in a compact building with the same floor space.

2 Usual ranges for A/V$_e$ ratios in differing residential building categories. 50% of buildings fall into the average range, while buildings with lesser or greater degrees of compactness are in the marginal ranges with 25% each. The variations within categories may appear small, but a house with multiple dwellings with a A/V$_e$ ratio of 0.6 m²/m³, therefore with low compactness, has an approximately 50% greater heat conducting envelope surface than a building in the same category with a A/V$_e$ ratio of 0.4 m²/m³, therefore with an average degree of compactness. According to that more heat energy is lost through the building envelope.

3 Average annual irradiation on differently oriented building surfaces with 900 kWh/m²a global radiation (horizontal surface 100%).

OPENINGS

4 Heating requirement of a room in a residential unit depending on the proportion of window area, orientation and heat transfer coefficient of the walls and windows. Here we are considering windows with a U-value of 1.6 or 1.0 W/m²K and oriented to the north, east/west, and south, built into walls with U-values of 0.4 or 0.2 W/m²K.

a External wall U-value 0.4 W/m²K. A windowless room has an annual heating requirement of approximately 35 kWh/m²a. With good-quality (1.0 W/m²K U-value) windows facing south, solar gains will reduce the heating requirement. The gains and losses cancel each other out with poor-quality windows facing south. The same applies to good-quality windows facing east or west. With other orientations, the solar gains cannot compensate for losses.

b With a superior quality of outer wall (0.2 W/m²K U-value) the heating requirement of a windowless room is approx. 25 kWh/m²a. With good-quality (1.0 W/m²K U-value) windows facing south, solar gains will reduce the heating requirement. With other orientations or poor-quality windows, the solar gains cannot compensate for losses. A larger amount of window surfaces therefore increases the heating requirement of the building.

——— Uw = 1.6 W/(m²K), South
— · — Uw = 1.6 W/(m²K), East/West
·········· Uw = 1.6 W/(m²K), North

——— Uw = 1.0 W/(m²K), South
— · — Uw = 1.0 W/(m²K), East/West
·········· Uw = 1.0 W/(m²K), North

5 Building volume design and alignment in relation to climate conditions and latitude.

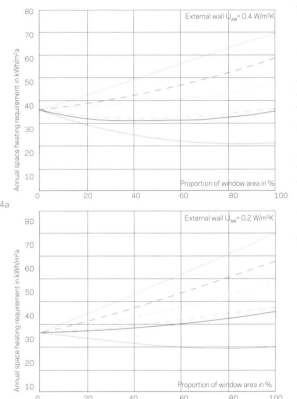

External wall U$_{AW}$ = 0.4 W/m²K
Proportion of window area in %
Annual space heating requirement in kWh/m²a

4a

External wall U$_{AW}$ = 0.2 W/m²K
Proportion of window area in %
Annual space heating requirement in kWh/m²a

b

The number and size of openings in the building envelope influence the building's energy balance due to two opposing effects. Despite improvements in design and materials in recent years, windows lose 2 to 5 times as much heat by transmission compared to opaque surfaces of the same size. A large proportion of window area therefore increases the heating requirement of the building. Yet windows reduce the need for heating by making use of solar gain. Solving these conflicting objectives means balancing heat losses and gains. This is influenced by orientation, proportion of window area, and the heat transfer coefficient of the walls and windows. ➘ 4 Improving the building envelope's insulation is more effective for reducing the heating requirement than the solar energy gained by having more glazing. With good thermal insulation, only good quality, south-facing (≤1.0 W/m²K U-value) windows will help to reduce the heating requirement. Conversely, this means that very large south-facing window areas do not increase the need for heating, but it is necessary to provide protection against overheating in the summer. Assuming the walls and windows are of good quality, east and west-facing windows will not significantly increase the heating requirement if the proportion of window area does not exceed 40%. Therefore low-energy houses with large windows must not necessarily face south. North-facing window areas need to be as small as possible. There is generally no need for solar screening here; on the other hand, losses will not be reduced or balanced by solar gains. ➘ 5

	Cold	Moderate	Dry	Tropical
Latitude (city)	60° north (Oslo)	50° north (Frankfurt am Main)	30° north (Cairo)	5° south (Kinshasa)
Building volume	compact, side-to-side ratio preferably 1:1:1	compact, oriented towards the sun	high storage mass, facing away from the sun	lightweight, well ventilated, facing away from the sun
Arrangement of main rooms	centrally in the building	centrally in the building or along the southern facade with a buffer zone towards the north	in the lower levels in order to limit heating up from solar irradiation	facing the main wind direction to benefit from natural ventilation
Arrangement of secondary rooms	surrounding the main rooms (onion principle – minimizing losses)	surrounding the main rooms or along the north facade as a buffer zone	in the top floors as buffer zone for the main rooms	corresponding to main rooms
Proportion of opening				
North	small	small	medium	high
East/west	medium	medium	small	small
South	moderate	high	small	high
Solar gains	desirable in winter and spring/autumn	desirable in winter; risk of overheating in spring/autumn	risk of overheating throughout the year; winter gains may be used for heating	not desirable; risk of overheating throughout the year
Use of solar energy/Arrangement of solar collectors on the building				
Support for heating	south facade	east/west facade	roof	not required
Domestic hot water	south facade/pitched roof	east/west facade; roof	roof	roof

5

ZONING BUILDINGS FOR OPTIMUM USE OF ENERGY

The layout of a building needs to be zoned to make optimum use of solar gains and to minimizing heat losses. In the layout of the building, rooms should be located depending on their function and requirements. ⤳ **1-3** For example, in the temperate climate of central Europe, storerooms, staircases, vestibules and rooms that are only used some of the time should face north in residential buildings. These zones serve as a buffer and reduce heat loss through the shaded north side of the building. Bedrooms and kitchens should face east, and living rooms and children's rooms should face south or west, in order to make use of the sun's rays to heat the building at the right times of day. In other climate zones (e.g. in desert climates) the sun's rays may result in overheating of buildings. Here zoning has to be applied differently. The main rooms will need to be situated on the side turned away from the sun.

There are also possibilities for energy-efficient vertical zoning of building layouts. By deliberately arranging differently heated zones, by using buffer zones and, depending on the orientation, by making use of passive solar gains it is possible to design for energy efficiency as well as achieve architecturally pleasing buildings. It is also necessary always to take account of the seasons and the hours of day and night. ⤳ **5**

In the case of apartment buildings, the heating requirements of individual dwellings vary considerably depending on the location in the building. Dwellings with many external surfaces have the greatest heating requirements, while those surrounded by heated spaces have the least. ⤳ **4**

Wherever possible, heating plant rooms should be situated on the top floor in the center of a building. The waste heat from this room can then be utilized to heat the surrounding living areas and the shorter pipe runs for heating and hot water systems reduce heat losses.

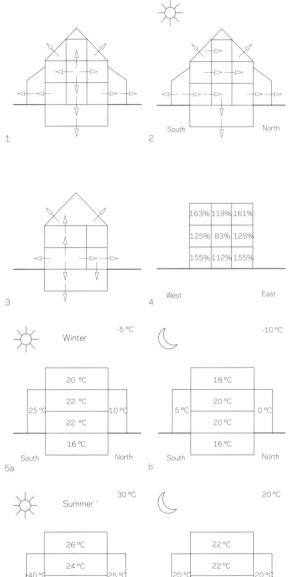

Design concepts for energy zoning of residential buildings:

1 Layered arrangement. The warmest rooms in the center of the building (living room, bathroom) are surrounded by rooms with low temperature levels (bedrooms, other ancillary rooms) and even unheated areas (conservatories). In this way, cooler rooms form a buffer zone around the warmer rooms and reduce their heat loss.

2 Sun-oriented arrangement. Living rooms face south. Heat losses through the envelope are compensated by solar gains. Ancillary rooms face north and so create a buffer for the areas of the building that do not receive any sun. This principle is often used in classic Passivhaus designs.

3 Staggered stories. The living rooms are on the south side of the ground floor, while the adjoining rooms and rooms that are only used part of the time (bedrooms) are above the living rooms and face north. This is a mixture of the first two principles.

4 Varying heating requirements of dwellings in apartment buildings depending on the position within the building.

5 Impact and temperature behavior of buffer zones depending on the orientation and the seasons as well as day and night. On sunny winter days, south-facing, unheated buffer zones can reach temperatures of up to 25 °C. At night and on the side away from the sun, the temperatures are lower and the conservatory acts as a buffer zone between the building and the outside. In particular, glass extensions can result in overheating of the building in summer. Rooms under ground remain at more or less the same temperature all the year round.

RESIDENTIAL BUILDING
This drawing shows the optimal arrangement of rooms in a residential building, taking into account their use as well as energy aspects. Dayrooms in constant use should be oriented to the south or should be surrounded by other rooms which are used temporarily at lower temperature levels (onion principle). Depending on the orientation of the rooms, different solar screening systems are suitable. → p. 84

a   Skylights facing north to provide daylight to internal rooms. Making use of the south-facing sloping surface for a photovoltaic/solar thermal installation.

b   Changing room/storage/WC facing north with minimal window areas.

c   Stairwell located towards the north as a buffer zone for the warmer dayrooms.

d   Bedrooms facing east (morning sun) with moderate amount of window area.

e   Bathroom (warmest room) surrounded by other rooms, with daylight from above, minimized heat loss.

f   Play and workroom facing west (afternoon sun) with moderate amount of window area.

g   Children's rooms facing south (receiving sunlight all day), moderate amount of window area, solar screening required.

h   Boiler room at a central location in the building; short pipe runs and losses heat up adjacent rooms.

i   Living and dining areas facing south (receiving sun all day), large amount of window area. Passive solar screening from cantilevered upper floor so that access to the terrace is unimpeded.

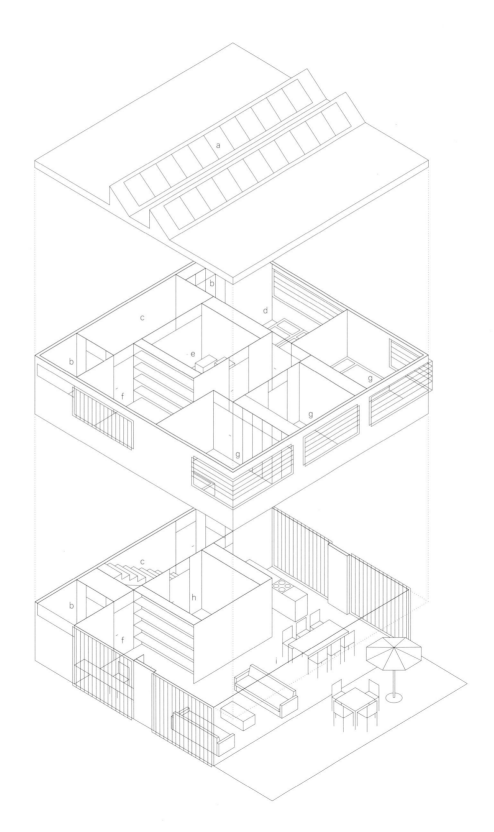

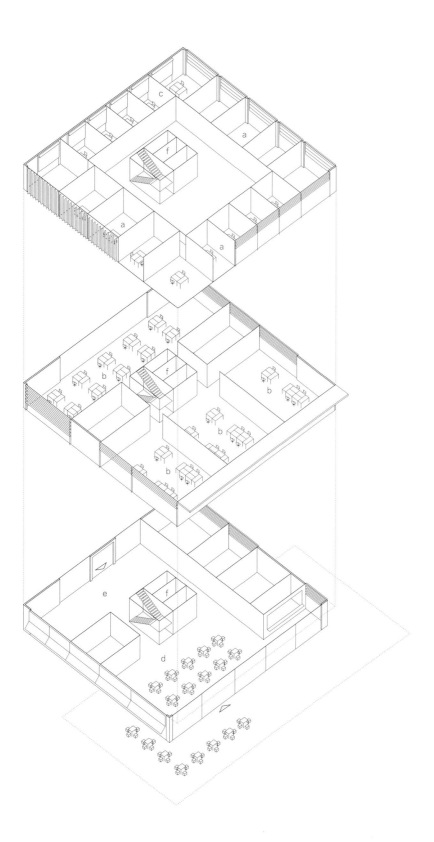

OFFICE BUILDING
Generally speaking, office buildings also have an optimal arrangement of rooms, depending on their use and taking into account energy aspects. The illustration shows solar screening systems for use on the east, west and south sides of the building. In the northern hemisphere glare protection is usually sufficient to the north.

a Cellular (individual) offices and meeting rooms with individually controlled solar screening and glare protection can be arranged on any side of the building.

b Group and open-plan offices should preferably face north or south and have rigid horizontal solar screening. This makes it possible to avoid user conflicts about artificial/natural lighting.

c Server rooms should be located on the north-facing facade so that excessive heat can be discharged naturally and quickly and so additional solar gain is avoided.

d Canteens benefit from access to a sunny open area facing east, south or west. If an associated outdoor area is to the east or west, the type of movable solar screening should be selected carefully to avoid conflicts in gaining access to the outside.

e Foyer and entrance to the north or south. Fixed solar screening on the south side avoids a conflict between necessary shading and opening the building.

f Rooms which need no natural daylight, such as printer rooms, kitchenettes, and storerooms, can be located at the interior of the building where they do not use up valuable facade space.

SCHOOL BUILDING
Depending on the use of the many types of rooms in school buildings and taking into account energy aspects and suitable solar screening systems, some general principles apply for optimizing the room layout in schools.

a   North-facing skylights to provide daylight for art classrooms. Making use of the south-facing sloping surface for a photovoltaic/solar thermal installation.

b   Art rooms facing north, with consistent daylighting. Internal glare protection.

c   Classrooms facing east, south, or west, with solar screening and glare protection controllable from each room; this may also be used for blacking out the room (e.g. external shutters).

d   Rooms which do not need natural daylight, such as WCs and storerooms, can be located at the interior of the building where they do not use up valuable facade space.

e   An atrium or inner courtyard can be used as a break area and to provide natural lighting to corridors.

f   Foyer and entrance to the north or south. Fixed solar screening on the south side avoids a conflict between necessary shading and opening the building.

g   Canteens/refectories benefit from access to a sunny open area facing east, south or west. If the open area is to the east or west, it is possible, depending on the type of solar screening, that a conflict arises with regard to access to the open area.

h   Staff and administrative rooms with individually controlled solar screening and glare protection can be arranged on any side of the building.

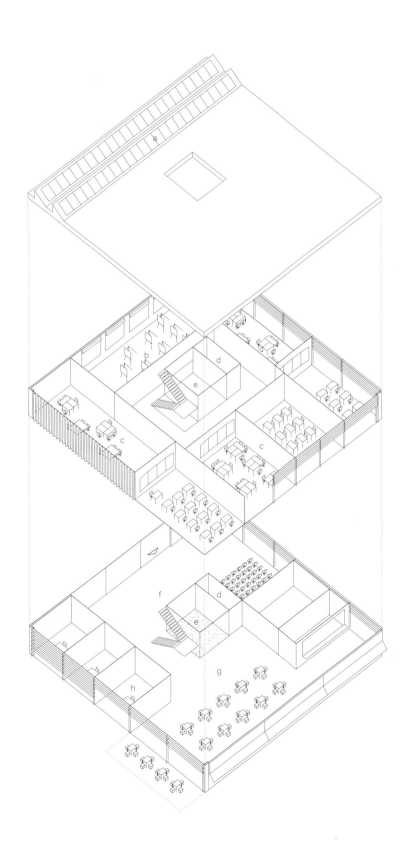

STORAGE CAPACITY / STORAGE MASS

Solar heat penetrates through the openings in a building and heats the interior. In order to be able to use this heat during hours of darkness and to reduce temperature fluctuations in a room, it needs to be stored somewhere. This is done by passive storage in the internal building components. A high degree of heat storage capacity increases the useful part of the passive solar gains and improves thermal comfort on hot days, without the need for technical equipment.

HEAT STORAGE

The storage capacity of a building material depends on its *density* (r) and its *specific heat capacity* (c), which allows us to determine its volume-based *heat storage value* (s). This expresses how much heat can be stored in a cubic meter of a specific building material. The ability of a building material to compensate for fluctuations in the room air temperature also depends on the speed with which it absorbs heat. This is influenced by its thermal conductivity. The thermal effect of a building material on the room air temperature can be expressed as the *heat penetration coefficient* (b) in W/m²K, which is the square root of the product of the density, heat storage value and thermal conductivity. It is necessary to remember that heat absorption decreases with time as the room temperature and building component temperature approach the same value. Heavy building component layers are particularly suitable for storing heat up to a depth of 8 to 10 cm. However, only the amount of heat that has been stored at temperatures in excess of the minimum desired room temperature is usable. This means that this use of solar energy becomes more efficient the greater the permissible fluctuation in the room temperature. This is only possible to a limited extent for workplaces in continuous use. When calculating the annual heating requirement Q_H according to DIN 4108-6, the effective storage capacity C_{eff} is determined from the heat storage capacity of the building envelope and interior building components in a building. The designer can assume a fixed C_{eff} of 50 Wh/(m³K) × V_e for heavy buildings and 15 Wh/(m³K) × V_e for lightweight buildings.

PHASE CHANGE MATERIAL (PCM)

The heat storage capacity of a building material depends among other things on its mass. However, if it is possible to make use of the phase change of a material, such as the transition from solid to liquid, then it is possible to achieve a high storage capacity largely independent of the mass. Materials that can make use of phase change to store heat are termed *phase change materials* (PCM). When a phase change occurs, additional heat energy is absorbed (latent heat) in order to alter the aggregate condition of a building material. There is no further rise in temperature of the building material until the phase change has been completed. The quantity of heat absorbed is relatively large in relation to the specific heat capacity, thus creating an exceptionally efficient heat store. For example, the amount of additional energy required to melt wax is equivalent to the energy required to raise the temperature of solid wax by 60 K. The melting point of a PCM should ideally be close to the maximum comfortable temperature of an interior. The phase shift is triggered when the air in the room reaches the melting temperature; the necessary heat is taken from the air in the room and a further rise in temperature is prevented. Two material groups with melting points between 20 and 30 °C are suitable as PCMs: organic PCMs (long-chain hydrocarbons, e.g. paraffins) and salt hydrates. Especially in lightweight buildings, PCMs are effectively used to increase the level of comfort in the summer, as these buildings usually lack other storage masses. ⟶1 This effect is however not taken into consideration in the usual verification procedures for heat protection.

Specific heat capacity (c) indicates in Wh/kg the amount of heat needed to heat a kilogram of a substance by 1 K, making it related to mass. Building materials can be classified into three types according to their specific heat capacities: metallic substances (hard non-iron metals excluding aluminum) c - 0.12, mineral substances (e.g. concrete) c - 0.28, and organic substances (e.g. wood) c - 0.58.

The heat storage value (s) indicates in Wh/m³ the amount of heat needed to heat a cubic meter of a substance by 1 K. The heat storage value is therefore based on the volume. It is calculated using the specific heat capacity times the density of a substance.

The heat penetration value (b) is the square root of the product of the density (r), the specific heat capacity (c) and the thermal conductivity (l). b = √(r × c × l).

Effective storage capacity C_{eff} is the sum of the products of the specific heat storage capacities (c), the densities (ρ), the effective layer thicknesses (d), and the room-side surface areas of the building elements (A).
$C_{eff} = \Sigma(c_i × \rho_i × d_i × A_i)$.
Here i includes all layers with a layer thickness up to 10 cm total thickness that are not separated from the room by insulation (λ < 0.1 W/mK and R ≥ 0.25 m²K/W). For internal building elements, half of their effective heat storage capacity is assigned to each room.

1 Example of temperature development (over a week) in a classroom of lightweight or solid construction, in each case with and without ventilation. Whereas the temperature only rises slowly in a solid construction, lightweight construction heats up considerably. But the heat can be rapidly dissipated at night through ventilation. The use of PCM makes it easier to keep constant temperatures in buildings with lightweight construction.

2 The storage capacity of a building material depends on its mass and the component's surface area. The chart shows the surface areas of different wall materials needed to store 1 kWh of cooling load when temperatures rise by 4 K (from 21 to 25 °C), therefore prevent the room from heating up. The mass and penetration depth are also given (d here is not the actual building element thickness).

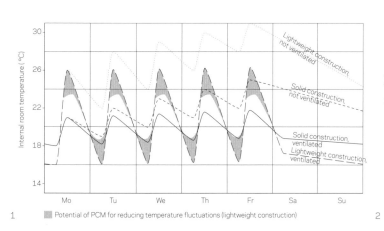

1 Potential of PCM for reducing temperature fluctuations (lightweight construction)

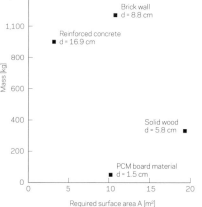

2 Required surface area A [m²]

HEAT TRANSMISSION / PHASE SHIFT

Thermal conductivity (λ) indicates the heat flow in watts that is transferred in a building element layer with a surface area of 1 m² and a thickness of 1 m at a temperature difference of 1 Kelvin. λ = W/mK.

Thermal resistance (R) is the quotient of the layer thickness (d) and the thermal conductivity (λ). R = d/λ.

Total thermal resistance (R_T) is calculated from the thermal resistances of the building elements (R) as well as from the internal and external transmission of heat (R_{si}, R_{se}) in m²K/W. $R_T = R_{se}$ + R1 + R2 + ... + RN + R_{si}.

Thermal transmittance (U-value) is the reciprocal value of the heat transfer resistance, stated in W/(m²K). The U-value was formerly called k-value in Germany. $U = 1/R_T$

3 Schematic presentation of the temperature amplitude values (TAV) and the phase displacement (η). The temperature amplitudes in each case show the temperature variations depending on the time of day internally and externally. The internal temperature variation lags behind the external temperature variation by what is called phase shift (η).

4 Internal and external heat transfer resistances depending on the orientation of the building element.

5, 6 Rauch House, Schlins, 2007, Roger Boltshauser and Martin Rauch. 85% of the building was constructed using material excavated on site. Very little energy was required for the construction. The thick exposed walls, made using clay from the excavation, have a high heat and moisture storage capacity and create a comfortable room climate.

MOISTURE STORAGE

The hygroscopicity of building materials, in particular those of the interior (walls, ceilings, and floor coverings, but also furniture), stores the moisture in the room. Materials that absorb a large amount of moisture from the air and can give it off again rapidly (mainly plant, animal, and porous mineral substances) are more useful than non-absorbent building materials (plastics, concrete, metals), as their room-side surfaces act as humidity buffers. While this has an effect on energy conservation, the main benefit of moisture storage is increased room comfort.

HEAT TRANSMISSION

If there is a temperature difference between the building's interior and the exterior air, energy will be conducted from the warm to the cold side by transmission through the building envelope. Depending on the thermal conductivity (λ) of the building materials used, differing amounts of energy will be lost through a building component. The lower the thermal conductivity (λ), the better is the heat protection offered by a building material of the same layer thickness. The heat protection effect is expressed as thermal resistance (R). The thermal resistance of a wall is calculated by computing the resistances of the layers of the building component. Still or slow-moving layers of air in building components also have a thermal resistance. Air layers in back-ventilated facades are however considered as exterior air.

The transmission of the air temperature to the exterior building components also provides resistance, termed the *heat transfer resistance* R_s. These values vary for the interior and exterior sides of building components, and also depending on the orientation (upwards, horizontal, downwards).

The insulation effect of a building component involves both the thermal resistances (R) of the individual building component layers and the heat transfer resistances

(R_{si}, R_{se}) on the interior and exterior sides. The total thermal resistance (R_T) therefore refers to the total resistance that the envelope presents to the loss of heat. In many countries this value is stated for the heat protection provided by building components; the aim is generally to achieve the highest possible value. In Germany, on the other hand, the thermal transmittance (U-value) is the calculation measure. The U-value simply describes the reciprocal value of R_T and the objective is to achieve low values for the building components in question.

PHASE SHIFT

The components of the building envelope are exposed to varying exterior temperatures during the course of the day. The components are heated and cooled to a 24-hour rhythm, which impacts the building component's thermal behavior and the interior temperature. Outdoor temperature fluctuations are tempered and transmitted with a delay to the interior. This delayed transfer to the interior is expressed as a phase shift in hours. The temperature amplitude ratio (TAR) shows the ratio between the maximum temperature fluctuation at the outer and inner surfaces of a building component and indicates the insulating effect of a building component. The smaller the TAR, the better the insulation offered by that building component and the greater the phase shift. A low temperature amplitude ratio is desirable especially for heat protection in summer, with an ideal phase shift of approximately 12 hours. The heat of a hot summer day will in that case be released into the interior with a delay of 12 hours and can, for example, be extracted from the building using cool night air. The temperature amplitude ratio depends on the heat storage value (s), the thermal conductivity (λ) and the thickness of the building component. A high heat storage value, low thermal conductivity, and greater building component thickness will have a positive effect on the temperature amplitude ratio. ⌐ 3

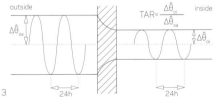

3

$$TAR = \frac{\Delta\hat{\theta}_{oi}}{\Delta\hat{\theta}_{oa}}$$

Direction of heat flow	upwards	horitzontal	downwards
R_{si} (internal)	0.10 m²K/W	0.13 m²K/W	0.17 m²K/W
R_{se} (external)	0.04 m²K/W	0.04 m²K/W	0.04 m²K/W

4

5

6

CONSTRUCTION METHODS

Lightweight and solid buildings behave differently on the basis of the building material properties described above. Lightweight buildings are generally subject to greater temperature and humidity fluctuations, and the resulting so-called barracks climate is felt to be uncomfortable. There is a risk of overheating, but at least such buildings can be quickly heated up and cooled down (for example by nocturnal cross-ventilation). This effect can be reduced by the use of PCMs. ➘ 1

The reduction in the annual heating requirement due to effective heat storage in internal building components varies approx. 8% between lightweight and solid construction. ➘ 2 Compared to thermal insulation, the storage mass of a building is thus of little importance to its heating energy requirement. Storage of irradiated solar heat moreover requires that the heating output can be quickly reduced when room temperatures are raised by solar irradiation. This is only possible to a limited extent with radiant heating systems (underfloor heating). There is a particular risk of overheating during transitional periods. With a well-insulated building envelope and a window area exceeding 30%, overheating during the warm season can only be prevented by the use of solar screening.

AIRTIGHTNESS

As the insulation standard of the building envelope increases, most of the heat is no longer lost due to transmission through the building components, but instead due to leakage in the building envelope. The Energy Conservation Regulations therefore require that the thermal building envelope should be durably airtight. With solid walls, the required airtightness is generally achieved by means of continuous plaster. In the case of lightweight building components, the vapor barrier is adhered to be airtight. The edge seals and joints in the different building components are particularly important regardless of the construction method used.

GRAY ENERGY IN BUILDING MATERIALS

Along with the energy a building consumes for heating, cooling, ventilation, and hot water provision during its use, energy is also consumed in the production, maintenance, and disposal of building components. This gray energy can be determined with a life cycle assessment of the building materials used, and it is expressed in MJ of primary energy. As energy consumption in the use phase is optimized further, gray energy becomes more important in the building design. The gray energy of a building material depends on the energy expended to extract raw materials, for manufacture, and for disposal options (recycling, incineration, landfill). As a function of a building's service life, durable building components are the most favorable as regards gray energy. Lightweight buildings generally require less primary energy in their construction. A building with a high degree of compactness also uses less gray energy because of its smaller facade. ➘ 3

The air change in a building, which measures the lack of airtightness of a building construction when the windows and doors are closed, is defined as one complete air change in the building per hour with a pressure difference of 50 Pa between the inside and outside.
Maximum permissible volume-based air change n_{50} according to EnEV 2009 for new buildings: for buildings with heating, ventilation, and air conditioning systems this is $n_{50} \leq 1.5$ h^{-1}. This means that, with a pressure difference of 50 Pa, no more than 1.5 times the volume of air in the whole building should escape through joints and cracks during one hour.
In the case of buildings with air conditioning systems this value is $n_{50} \leq 3.0$ h^{-1}. This can be checked by means of a blower door test.

1 Qualitative evaluation of the thermal properties and the primary energy consumed for different building methods.

2 Impact of heat storage capacity on the annual heating requirement of a single family house of solid and lightweight construction.

3 Comparison of the energy required to construct a building using solid and lightweight construction methods and for compact and non-compact buildings.

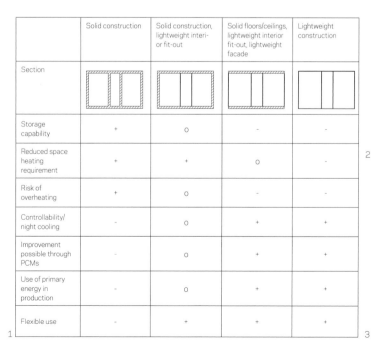

	Solid construction	Solid construction, lightweight interior fit-out	Solid floors/ceilings, lightweight interior fit-out, lightweight facade	Lightweight construction
Section				
Storage capability	+	o	−	−
Reduced space heating requirement	+	+	o	−
Risk of overheating	+	o	−	−
Controllability/ night cooling	−	o	+	+
Improvement possible through PCMs	−	o	+	+
Use of primary energy in production	−	o	+	+
Flexible use	−	+	+	+

1

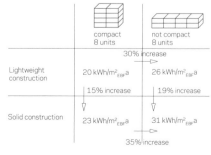

2

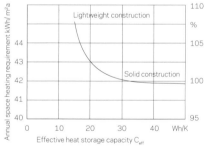

3

OPAQUE BUILDING COMPONENTS / EXTERNAL WALL CONSTRUCTIONS

Opaque external wall construction for refurbishing buildings subject to a preservation order. This external wall construction does not comply with the requirements of the EnEV 2009, but this type of interior insulation is still used for the refurbishment of buildings. The thickness of the internal insulation is determined by the space available. Thermal insulation inside a building takes up valuable floor space.

4 Internal insulation with calcium silicate panels.

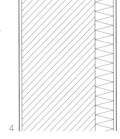

Building material	Layer thickness [cm]	Density [kg/m³]	Total PEI [MJ/m²]
Lime cement plaster	1.5	existing	existing
Brickwork	40	existing	existing
Calcium silicate panels	10	300	666
Lime cement	1.5	1,400	31
Total	53		697

Service life of external skin: not known
Ease of dismantling: o

$C_{eff,i}$: 31 Wh/K
Temperature amplitude ratio: 0.021
Phase shift [h]: 14

Opaque external wall constructions with a thermal transmittance (U-value) of 0.28 W/m²K. This is equivalent to the building element quality for refurbishment installations.

5 Core insulation. Walls with this type of construction display a high level of heat storage capacity and durability, yet they require a great deal of energy to build (gray energy).

6 Monolithic wall (bricks). It is difficult to fulfill the ever-increasing thermal insulation performance requirements using monolithic walls. Such walls display good durability and average heat storage capacity. The disposal of this type of construction is possible with a hardcore crushing process and without difficulties in separating materials.

7 Monolithic wall (insulating concrete). Such a wall will need to be 70 cm thick to meet the requirements of EnEV 2009 for external walls. As with brick walls, this insulating concrete wall is extremely durable and is easily recycled.

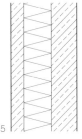

Building material	Layer thickness [cm]	Density [kg/m³]	Total PEI [MJ/m²]
Reinforced concrete	8	2,400	238
Polyurethane insulation	10	30	263
Reinforced concrete	17.5	2,400	522
Total	35.5		1,023

Service life of external skin: 80 yrs.
Ease of dismantling: -

$C_{eff,i}$: 67 Wh/K
Temperature amplitude ratio: 0.019
Phase shift [h]: 5.8

Building material	Layer thickness [cm]	Density [kg/m³]	Total PEI [MJ/m²]
Lime cement plaster	2	1,800	41
Poroton block	36.5	740	449
Gypsum plaster	1.5	1,200	35
Total	40		525

Service life of external skin: 45 yrs.
Ease of dismantling: o

$C_{eff,i}$: 26 Wh/K
Temperature amplitude ratio: 0.001
Phase shift [h]: 25.2

Building material	Layer thickness [cm]	Density [kg/m³]	Total PEI [MJ/m²]
Insulating concrete	70	800	865
Total	70		865

Service life of external skin: 60 yrs.
Ease of dismantling: +

$C_{eff,i}$: 14 Wh/K
Temperature amplitude ratio: 0.001
Phase shift [h]: 26.8

Building material	Layer thickness [cm]	Density [kg/m³]	Total PEI [MJ/m²]
Synthetic resin plaster	2	1,100	399
Expanded polystyrene	16	25	304
Calcium silicate block	17.5	1,800	310
Gypsum plaster	1.5	1,200	35
Total	37		1,048

Service life of external skin: 40 yrs.
Ease of dismantling: -

$C_{eff,i}$: 48 Wh/K
Temperature amplitude ratio: 0.024
Phase shift [h]: 4.2

Opaque external wall constructions with a thermal transmittance (U-value) of 0.24 W/m²K. This is equivalent to the building element quality for EnEV reference buildings.

1 Thermal insulation sandwich system. This construction method is popular and cost-effective. It combines the good storage characteristics of brickwork with the good insulation characteristics of expanded polystyrene (EPS). Negative aspects however are the relatively limited durability and the difficulty of separating the individual layers, making recycling expensive.

Building material	Layer thickness [cm]	Density [kg/m³]	Total PEI [MJ/m²]
Fiber cement board	0.8	1,700	250
Aluminum profile	18	2,800	341
Mineral wool	16	46	150
Brickwork	17.5	1,800	1,100
Clay plaster	1.5	1,500	1
Total	37.8		1,842

Service life of external skin: 55 yrs.
Ease of dismantling: o

$C_{eff,i}$: 49 Wh/K
Temperature amplitude ratio: 0.031
Phase shift [h]: 6.5

2 Back ventilated facade with fiber cement panels. As regards its thermal properties, this type of wall construction is comparable to the thermal insulation sandwich system above, although it is more durable and easier to dismantle.

3 Wood facade with softboard as a substrate for plaster.

4 Wood facade with lap siding. Lightweight wooden facades only store a small amount of heat, yet less primary energy is expended in their manufacture than with solid constructions.

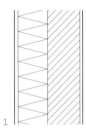

Building material	Layer thickness [cm]	Density [kg/m³]	Total PEI [MJ/m²]
Mineral plaster	0.5	1,400	13
Wood fiber insulation board	6	350	851
Fiber insulation material	16	46	129
Timber studs	16	600	196
OSB (oriented strand board)	2	650	267
Battens / counter battens	4.8	600	61
Gypsum plasterboard	1.25	900	40
Total	30.55		1,557

Service life of external skin: 30 yrs.
Ease of dismantling: o

$C_{eff,i}$: 3 Wh/K

Area 1 (timber supports):
Temperature amplitude ratio: 0.021
Phase shift [h]: 13.2

Area 2 (insulation):
Temperature amplitude ratio: 0.022
Phase shift [h]: 1.8

Building material	Layer thickness [cm]	Density [kg/m³]	Total PEI [MJ/m²]
Lap siding	2.2	600	195
Battens	2.4	600	30
Bitumen-impregnated fiberboard	2	250	203
Loose cellulose insulation	14	50	-55
Timber studs	14	600	171
OSB (oriented strand board)	2	650	267
Fiber insulation material	4.8	46	22
Battens/counter battens	4.8	600	61
Gypsum plasterboard	1.25	900	40
Total	28.65		934

Service life of external skin: 50 yrs.
Ease of dismantling: +

$C_{eff,i}$: 3 Wh/K

Area 1 (timber supports):
Temperature amplitude ratio: 0.022
Phase shift [h]: 13.1

Area 2 (insulation):
Temperature amplitude ratio: 0.019
Phase shift [h]: 1.6

OPAQUE BUILDING COMPONENTS / ROOF CONSTRUCTIONS

Roof constructions with a thermal transmittance (U-value) of 0.20 W/m²K. This is equivalent to the building element quality for EnEV reference buildings.

5 Tiled roof.

6 Green roof.

7 Metal sheet roof (back ventilated).

8 Lightweight flat roof.

5

Building material	Layer thickness [cm]	Density [kg/m³]	PEI [MJ/m²]
Roof tiles	1.3	1,800	149
Counter battens	2.4	600	20
Battens	2.4	600	20
Underlay	0.02	920	27
Wood fiber insulation board	10	350	2,839
Rafters	14	600	176
Fiber insulation material	14	115	156
PE vapor barrier	0.02	920	14
Gypsum plasterboard	1.25	900	40
Total	31.4		3,441

Service life of external skin: 60 yrs.
Ease of dismantling: +

$C_{eff, i}$: 7 Wh/K

Area 1 (rafters):
Temperature amplitude ratio: 0.026
Phase shift [h]: 12.3

Area 2 (insulation):
Temperature amplitude ratio: 0.023
Phase shift [h]: 2.3

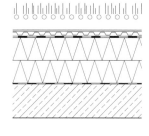

6

Building material	Layer thickness [cm]	Density [kg/m³]	PEI [MJ/m²]
Vegetation layer	8	-	-
Filter fleece	0.5	1,140	183
Drainage layer	2.5	600	73
EPDM roofing membrane	0.15	1,200	362
Tapered EPS insulation	24	25	935
PE vapor barrier	0.02	950	23
Reinforced concrete	18	2,430	654
Total	53.2		2,230

Service life of external skin: 30 yrs.
Ease of dismantling: o

$C_{eff, i}$: 3 Wh/K

Temperature amplitude ratio: 0.022
Phase shift [h]: 1.7

Area 2 (insulation):
Temperature amplitude ratio: 0.024
Phase shift [h]: 6.0

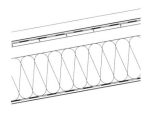

7

Building material	Layer thickness [cm]	Density [kg/m³]	PEI [MJ/m²]
Steel sheet (galvanized)	0.06	7,800	150
Separating membrane	0.1	950	120
Tongue-and-groove siding	1.6	600	266
Rafters	22	600	275
Mineral fiber insulation	22	45	249
Vapor barrier	0.02	950	14
Gypsum plasterboard	1.25	900	40
Total	25.3		1,114

Service life of external skin: 40 yrs.
Ease of dismantling: o

$C_{eff, i}$: 7 Wh/K

Temperature amplitude ratio: 0.021
Phase shift [h]: 1.6

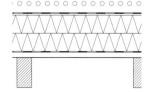

8

Building material	Layer thickness [cm]	Density [kg/m³]	PEI [MJ/m²]
Gravel	5	2,500	14
Roofing membrane	0.15	1,200	362
Tapered EPS insulation	20	25	760
Vapor barrier	0.02	950	23
Tongue-and-groove siding	2	600	169
Rafters	8	600	100
Total	35.2		1,428

Service life of external skin: 30 yrs.
Ease of dismantling: o

$C_{eff, i}$: 48 Wh/K

Temperature amplitude ratio: 0.021
Phase shift [h]: 1.6

THERMAL BRIDGES

The term *thermal bridge* refers to thermal weak points in the insulated building envelope, where there is increased transmission heat loss from inside to outside in comparison to the rest of the building envelope. ⤹ 4 Thermal bridges increase the heating requirement of the building. In particular, in the case of well-insulated buildings, thermal bridges can constitute up to 50% of the entire transmission heat loss. The increased cooling in the interior may be experienced as uncomfortable. Furthermore, condensation may form in the area of the thermal bridges. This may result in damage from moisture and cause mold to grow.

The behavior of thermal bridges can be represented using heat flow lines and the temperature fluctuation in a building component. ⤹ 2 Various forms of thermal bridges are distinguished by their physical causes; often various aspects interact with each other. ⤹ 1

GEOMETRIC THERMAL BRIDGES
Geometric thermal bridges always appear where there is a large exterior surface corresponding with a considerably smaller inner surface due to the shape of the building. This is the case, for example, with external corners of buildings, the roof ridge, dormer and bay windows, eaves, verges, and near the ground. It is possible to reduce the number of geometric thermal bridges by adopting a compact building shape.

STRUCTURAL THERMAL BRIDGES
Structural thermal bridges are areas in a building envelope where, mainly for construction reasons, building materials are used that transmit heat better or there is less insulation than surrounding areas. The increased heat conductivity in this area causes more heat to be lost around these materials. Structural thermal bridges often occur in skeleton constructions in the area of the load-bearing elements (e.g. pillars and rafters) in the building envelope. However, window frames, poorly insulated shutter boxes and lintels over windows as well as wall/base connections and floor coverings in exterior walls are structural thermal bridges. Structural thermal bridges can be avoided in the design by strict separation of the functions of the load-bearing structure and the shell.

CONVECTIVE THERMAL BRIDGES / MASS FLOW
These thermal bridges arise where a flowing medium results in increased disposal of heat in or on a building component. This mainly concerns air flows, which may cause cooling of adjacent areas when there are leakages in the building envelope. Theoretically, water and drainage pipes installed in walls may also create mass flow-based thermal bridges. Such thermal bridges can be avoided by means of careful construction methods.

ENVIRONMENT-BASED THERMAL BRIDGES
These thermal bridges arise when the heat transmitting behavior of exterior building components is influenced by heating elements or furnishings. Radiators along an exterior wall cause it to heat more on the interior, resulting in increased heat flow through the building component. Conversely, the arrangement of furniture and furnishings (e.g. curtains) can reduce the heat exchange between a building component and the room air in cases where air cannot adequately circulate behind. This leads to increased cooling of the components of the building envelope, which can result in condensation and the formation of mold. Environmentally determined thermal bridges are partly dependent on the behavior of the user, and cannot therefore be entirely excluded at the design stage, but designers need to know the interrelationships and explain these to the user.

THERMAL BRIDGE CALCULATION AND THERMAL BRIDGE CATALOG
In order to assess thermal bridges, the minimum surface temperature on the interior of a building component and the additional transmission heat loss through the thermal bridge can be compared with an undisturbed building component surface. Because of the complex interactions and the non-linear movement of the heat flow in a building component, the calculation of thermal bridges is complicated and is performed using special programs that require extensive training and precise knowledge of the physical interrelationships. For this reason thermal bridge catalogues have been made available to designers. These include the results of systematic calculations for many types of thermal bridges. ⤹ 3

THERMAL BRIDGES AND THERMAL INSULATION CERTIFICATION
The Energy Conservation Regulations (EnEV) provide for three different detailed processes in order to account for thermal bridges:
- If no details of thermal bridges are recorded, then the effects are calculated by adding a fixed figure of 0.10 W/m²K to the average thermal transmittance values of all the components of the building envelope.
- If the designers have documented the thermal bridges and their effect is reduced in accordance with the requirements of DIN 4108 Bbl 2, then a fixed figure of 0.05 W/m²K will be added.
- If the thermal bridges have been calculated precisely, then it is usually possible to add the lower figures.
The minimum thermal insulation of building components as required by DIN 4108-2 is also to be provided in the area of thermal bridges.

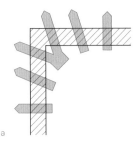

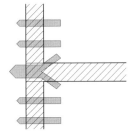

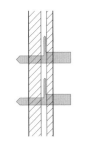

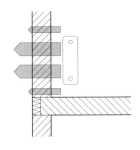

1 Types of thermal bridges.
a Geometric thermal bridges (e.g. external wall corner).

b Structural thermal bridges (e.g. floor supports).

c Mass flow-based thermal bridges (e.g. gaps in the building envelope).

d Environment-based thermal bridges (e.g. radiators next to an external wall).

2 Presentation of thermal behavior of a geometric thermal bridge based on an exemplary external wall corner, using heat flow lines and the temperature variations in the building component. On the room side, the distance between heat flow lines represents the width of wall one meter in height through which 1 watt of heat flows. In undisturbed wall areas, the distance between heat flow lines is constant; they run vertically through the wall and so cut across the temperature variation lines, which show the drop in temperature from inside to outside within the wall. The heat flow is increased in the area of thermal bridges, the heat flow lines are closer together; therefore more heat is transmitted per square meter of wall. The temperature variation lines also show a drop in temperature in the building component near the thermal bridge. As a consequence the wall temperature drops on the room side of the thermal bridge and there is a risk of condensation forming in the corner.

3 Example from a thermal bridge catalog for a window reveal joint with double-skin brickwork. The thermal bridge is situated near the external brick skin where it turns inwards.
The thermal bridge loss coefficient from the insulation material thickness (a) can be estimated from the table. This figure is multiplied by the length of the thermal bridge (in this case the height of the window opening) and the difference between the interior and exterior temperature to give the heat flow through the building element in watts.
It is possible to calculate the minimum surface temperature on the inside of the building element using the temperature factor f_{RSi} as shown. To do this, the temperature factor is added to the exterior temperature and multiplied by the difference between the interior and exterior temperature.
In both cases, unfavorable values between -10 and -15 °C are assumed for the exterior temperatures.

4 Examples of significant thermal bridges in a residential building

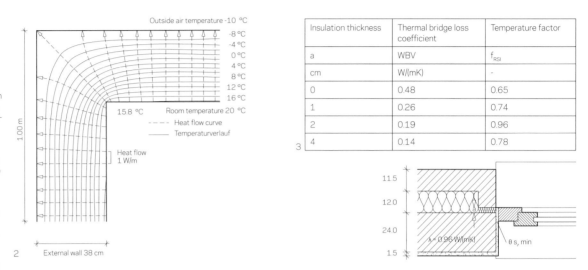

2

Insulation thickness	Thermal bridge loss coefficient	Temperature factor
a	WBV	f_{RSi}
cm	W/(mK)	-
0	0.48	0.65
1	0.26	0.74
2	0.19	0.96
4	0.14	0.78

3

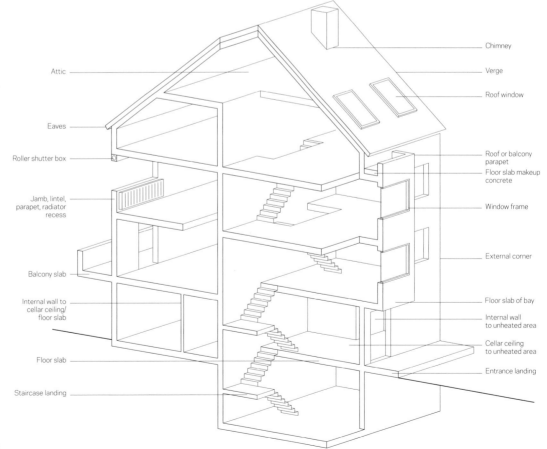

4

TRANSPARENT BUILDING COMPONENTS

The percentage of openings in a building envelope has an effect on the energy used in a building. ⌐ p. 69 Moreover the type and the quality of the windows used are also important. Transparent building components provide daylight to a room, a visual connection, and can contribute to ventilation and provide protection from rain, wind, cold, and noise. It must be possible to close building components to make them airtight, like the rest of the building envelope.

HEAT GAINS AND LOSSES

The energy performance of windows is determined by the transmission loss and the realizable solar gains. The performance also depends on the season. ⌐ 2 Gains and losses from transparent building components are taken into account when looking at the energy balance of buildings (e.g. according to DIN 4108 or DIN V 18599).

The insulation effect of a transparent building component is expressed as the heat transfer coefficient or thermal transmittance (U-value), as with opaque building components. For windows, this is calculated as the U_w value (w = window) in accordance with DIN EN ISO 10077-1. ⌐ 3 Frames and glazing are considered separately. In addition, the window format and the distance from the frame's edge to the glass are included in the calculation. In general, glazing offers a better U-value than the frame. What is more, a great deal of heat is lost in the transition between the frame and the glazing. It is therefore best to minimize the glass-to-frame ratios, and to avoid sash bars (glazing bars dividing or inserted between glass panes). Typical U_w values for windows according to the Energy Conservation Regulations 2009 are between ≤ 1.3 and 1.7 W/(m²/K), and for Passivhaus buildings ≤ 0.8 W/(m²/K). The solar gains from a transparent building component are described by means of the energy transmittance total (g-value). The g-value states the proportion of solar energy striking the building that is transmitted by the window into a room. The majority of this energy arrives in the interior by means of direct radiation transmission, while a smaller portion of the energy is absorbed by the glass pane and then emitted into the room (secondary heat release). ⌐ 4 Nothing can be deduced about the brightness of the room from the g-value. For this the light transmittance TL determined in accordance with DIN EN 410 is used. For the energy efficiency of a window, the following applies: the better the U-value and therefore the heat protection, the worse the total energy transmittance (g) and therefore the solar gains.

WINDOW TYPOLOGIES

Depending on the arrangement of the movable sashes, window typologies can be classified as basic, linked, and counter-sash windows. ⌐ 5 Basic frame windows have one opening sash with single or multiple glazing. In the case of linked windows, two sashes are laid on top of each other in the frame, which can be opened together or singly (for cleaning purposes). The combination of several panes increases the heat and noise insulation. Counter-sash windows consist of two basic frames, which are linked in a single box with an intermediate space of 10 to 15 cm. With all window types the arrangement of several panes will result in a reduction of transmission heat losses.

GLAZING TYPES / COATINGS / FILL MATERIALS

Glass for use in windows is manufactured using the float process. The heat insulation properties of single glass panes with a thickness of 3 to 19 mm are too poor, however, to allow for their use in heated buildings (U-value 5–6 W/m²K).

The panes of glass are therefore joined using linear seals in the edge details to make double or multiple insulated glazing. The permanent layer of air in the 12–20 mm space between the window panes, the glazing cavity, significantly improves the insulation effect. The U-value of double glazing is approximately 3W/m²K, which is half the value for single glazing. Plastic spacers wrapped with stainless steel foil around the edge details are used, which themselves have a very low specific thermal conductivity (warm edge).

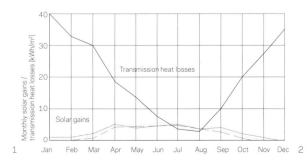

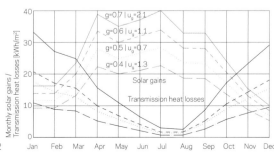

1

2

1, 2 Ratio of solar gains and transmission losses for transparent and opaque building elements during the year. Unlike opaque building elements, solar gains affect the energy balance of buildings in the case of transparent building elements. Except in the winter months, the solar gains exceed the transmission losses through windows, depending on the quality of the glazing. This may result in overheating of the building in summer. It should also be borne in mind that solar gains depend on the orientation of the window and the time of day.

3 Calculation of the thermal transmittance U_w

$UW = (A_f \times U_f + A_g \times U_g + I_{fg} \times \Psi_{fg}) / A_f + A_g$

U_w thermal transmittance of the entire window
U_g thermal transmittance of the glazing
U_f thermal transmittance of the frame
A_g glazing area
A_f frame area
Ψ_{fg} linear thermal transmittance frame – glass
I_{fg} length of edge frame/glass

4 Calculation of total energy transmittance (g-value).

$g = \tau_e + q_i = 36\%$

ρ_ϵ = reflectance
q_a = secondary heat release to the outside
q_i = secondary heat release to the inside
τ_e = direct radiation transmission

5 Window types
a Basic frame with single glazing
b Basic frame with double glazing
c Basic frame with triple glazing
d Linked sash window with 2 panes
e Linked sash window with 3 panes
f Counter sash window

6 Construction and technical parameters of various sorts of glazing. The emission value ϵ_n states how much radiation a body emits in comparison to an ideal heat radiator, namely a black body. Values approaching zero are ideal for a low-E coating.

The properties of double glazing can be improved further by coating the glass panes. Heat and sun protection coatings are particular useful from an energy-saving perspective. Silver-based low-emissivity coatings (lowercase-E) will improve heat protection. These selective coatings on the interior of the pane reflect a large part of the invisible infrared heat radiation, while at the same time the visible radiation ranges can pass through largely unhindered. Room heat is therefore reflected back into the room, while visible light is allowed to enter the room from outside. The radiation in the infrared range can therefore be reduced by up to 80% without the light transparency going below 0.7. Solar protection glazing made with silver, gold, or other metal oxides has the same effect. The incident solar radiation is reflected back, while the direct radiation transmission (τ_e) is reduced and the room heats up less. Effective solar protection is always linked to a large reduction in the transmission of light.

The space between the panes can be filled with argon or krypton gas in order to improve the insulating properties of insulation glass panels. These gases conduct heat very poorly. As a result, current glass pane standards can achieve U_w values as low as 1.1 to 0.7 W/m²K.

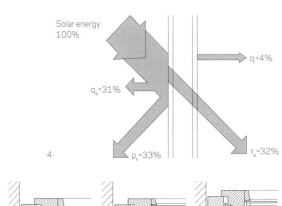

Solar energy 100%
$q_i=4\%$
$q_a=31\%$
$\rho_\epsilon=33\%$
$\tau_e=32\%$
4

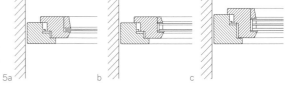

5a   b   c

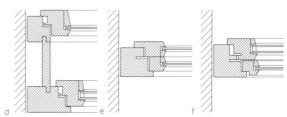

d   e   f

Ar Argon fill Kr Krypton fill LE Low emission layer ϵ_n SF Sound insulation foil	Inert gas Metal coating Sound insulation foil	Construction, glass thickness, glazing cavity	Overall thickness	Thermal transmittance U_g	Total energy transmittance g	Light transmission TL	Sound reduction index R_w
		mm per layer	mm	W/m²K	%	%	dB
Single glazing		6	6	5.6	85	90	
Double glazing		4/16/4	24	3	75	80	28–29
Heat protection glazing (Ar, LE ϵ_n= 0.1)		4/16/4	24	1.4	60	75	30–31
Heat protection glazing (Ar, LE ϵ_n= 0.03)		4/16/4	24	1.1	60	80	30–31
Heat protection glazing (Kr, LE ϵ_n= 0.03)		4/8/4/8/4	28	0.7	50	71	32
Sound insulating glazing (SF, Ar, LE ϵ_n= 0.03)		10/16/4/4	34	1.1	55	75	45
Sound insulating glazing (SF, Ar, LE ϵ_n= 0.03)		8/8/20/4/4	44	1.1	48	72	51
Solar glazing (Ar, LE ϵ_n= 0.03)		6/16/4	26	1.1	22	41	36

6

SHADING AND SOLAR PROTECTION

Thermal losses through the building envelope are offset by varying degrees of solar gains, depending on the season. In this regard the solar radiation available to the building does not correspond to the energy needs of the building at all times of the day. An excessive amount of solar radiation can lead to overheating of the building on sunny days, particularly in the case of well-insulated buildings. The use of screening systems can regulate the solar incidence to varying degrees.

Buildings in Germany have to be constructed so they meet the summer heat protection requirements according to DIN 4108-2. The result for residential buildings is that it is possible to do without active cooling and the associated energy consumption in the summer. In non-residential buildings, the need for cooling is reduced by the use of shading systems. Depending on the orientation of the window, proof of heat protection in summer is required if the window area is greater than 7–15% of the room floor area. ⟍1

SOLAR GAIN INDEX

The regulation for summer heat protection requires that the solar gain index S for critical spaces along the exterior facade, which are particularly exposed to solar radiation, has to be below the permitted solar gain index S_{per}. The solar gain index in a room is calculated on the basis of the window areas, the total energy transmittance (g-value) of the glazing including the solar protection and the room floor area. ⟍2

The effect of a sun screen is represented by a reduction factor F_C of the total energy transmittance (g-value) of the glazing. ⟍3 The F_C value depends on the type of sun screening ⟍5 and describes the effectiveness of the dif-

ferent systems for preventing overheating. The lower the F_C value, the more effective is the sun screen.

SUN SCREENING DEVICES

Sun screening devices can be applied to the outside of a building, on the inside, or in the space between the glass panes. It is also possible to install sun protection glazing. ⟍5 Solar protection should preferably not make the interior darker, to avoid the need for artificial lighting. External sun screens are most effective, because they shade the windows and offer the most effective protection against the room overheating. These may be rigid devices, protruding construction elements, or roller blinds, venetian blinds, or shutters, whether rolling or folding. Rigid sun screens will only offer sufficient protection when the sun is high if the screen extends far enough outside the window. For this reason they are mainly used on the south faces of buildings. They generally do not suffice as the only solar screening, but they can be combined with sun protection glazing without spoiling the building's appearance. It is also possible to respond more flexibly to changing irradiation and light conditions with movable installations. Therefore these are better for east or west faces. They may however significantly limit visibility and daylight (e.g. rolling shutters).

With sun screening devices inside the building, the solar radiation first enters the room and is then reflected or absorbed there, which contributes to a tempreature rise in the room. Interior sun protection is thus less effective than exterior sun protection, but this is offset by lower maintenance costs, as the solar protection is not exposed to the elements.

Calculation of the solar gain index S with window surfaces in room A_W in m², total energy transmittance g_{total} and net floor area of room A_G in m²
$$S = (A_W \times g_{total})/A_G$$

The total energy transmittance g_{total} is calculated from the g-value of the glazing and the reduction factor for sun screening F_c.
$$g_{total} = g \times F_c$$

1 Permissible values of floor area related window surface portion dependent on the orientation and incline of the windows, below which there is no need to provide proof of heat protection in summer according to DIN 4108-2, and climate regions for the calculation of the permissible solar gain index S_{per}.

2 Proportional solar gain indices S_x to determine the permissible solar gain index S_{per}.
The permitted solar gain index S_{per} is calculated taking into account the climatic region, construction type, mechanical ventilation as well as the window inclination and orientation. To arrive at this figure, the influencing factors S_x are added to S_{per}.

Tilt of window measured from horizontal	Orientation of windows		Window area as a proportion of floor area f_{AG} [%]
> 60° < 90°	From north-west via south to north-east	(compass diagram N/E/S/W)	10%
	North		15%
0° to 60°	All orientations		7%

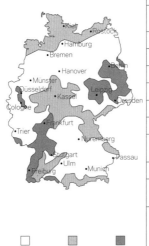

2   □ region A   ▨ region B   ▥ region C

Influencing factor	S_x
Climate region A/B/C	0.04/ 0.03/ 0.015
Form of construction/effective thermal storage capacity:	
Lightweight construction: without calculation of C_{eff}/A_G	0.06 f_{gew}
Medium construction: 50 Wh/Km² ≤ C_{eff}/A_G ≤ 130 Wh/Km²	0.10 f_{gew}
Heavy construction: C_{eff}/A_G > 130 Wh/Km²	0.115 f_{gew}
Night ventilation:	
with lightweight and medium construction/with heavy construction	0.02/ 0.03
alternatively: Solar glazing with g ≤ 0.4 or solar screening device to protect from diffuse radiation with g_{total} < 0.4	0.03
Window tilt: 0° ≤ tilt ≤ 60° (measured from horizontal)	-0.12 f_{neig}
Window orientation: windows facing north, northeast, and northwest with a tilt > 60° or windows with permanent shading from the building	0.10 f_{nord}
f_{sel} = window area + 0.3 × external wall area + 0.1 × roof and floor areas adjacent to unheated/net floor area f_{tilt} = tilted window area/net floor area f_{north} = window area facing north, northeast, and northwest/total window area	

1

3 Principle of reduction factor Fc

4 Suitable sun protection systems and combinations for different orientations. Rigid systems and horizontal slats are suitable for south-facing rooms. Vertical slats are more suitable for orientations to both the east and west. Outside awnings, roller blinds, and venetian blinds as well as sun protection glazing are suitable for all orientations, but vary in efficiency.

a Horizontal rigid sun screening, vertical slats on the interior (dazzle protection)
b Horizontal rigid sun screening, sun protection glass
c Horizontal slats
d Vertical slats
e Outside awning
f Sun protection glass

5 Various sun screening devices and their reduction factors F_c to reduce the total energy transmission as well as effects on the room and its use.

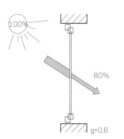
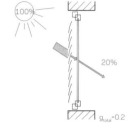

3 $F_c = g_{total} / g = 0.2 / 0.8 = 0.25$

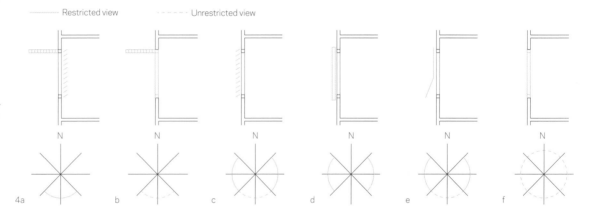

------- Restricted view   ----- Unrestricted view

4a   b   c   d   e   f

Device	None	Internal		Glazing cavity	External				
	no solar screening	inside roller blind	inside venetian blind	roller blind between glass panes	screen, canopy, loggia, balcony	outside roller blind	outside venetian blind	outside awning	roller shutter, folding shutter
Reduction factor F_c	1	0.8-0.9 0.7-0.8	0.75	0.8-0.9 0.7-0.8	0.5	0.4-0.5	0.25	0.4 0.5	0.3
Utilization of daylight	unrestricted	reduced	light redirection possible	light redirection possible	light redirection may be possible	reduced	light redirection possible	reduced	reduced
Glare protection	no	yes	depends on angle	depends on angle	no	yes	depends on angle	yes	yes
Transparency	unrestricted	depends on material	depends on angle	depends on system	unrestricted	depends on material	depends on angle	depends on system	restricted
Controllability	-	good	very good	depends on system	-	good	very good	very good	good
Maintenance requirement	-	low	low	low	-	high	high	high	medium

5

VENTILATION ELEMENTS IN THE BUILDING ENVELOPE

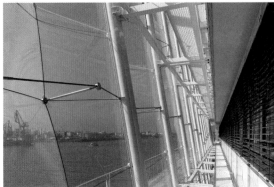

1   Window ventilation.

2   Unilever headquarters building, Hamburg, 2009, Behnisch Architects.
All workspaces have operable windows and outside solar screening, which is individually controllable. The daylight-optimized blinds are protected from strong wind and other weather influences by a single-layer film facade placed in front of the building's insulation glazing.

In order to guarantee the comfort of the interior of the building, the necessary minimum number of air changes appropriate for the use of the room or building must be achieved. ⤥ **chapter 3** Furthermore, high temperatures in buildings can be reduced by exchanging the air (e.g. during cool summer nights). The building envelope contributes to air changes by allowing air to pass through (uncontrolled) gaps and (controlled) openings. This natural air change is caused by the temperature difference between the interior and exterior, as well as by the wind. Temperature differences create vertical flows due to density differences. This works best with vertical opening formats and the air flow becomes more vigorous as the temperature difference increases. It is easier to ensure air change in winter than in summer. Wind contributes to natural air changes, partly from differing pressure conditions on the exterior walls and partly from turbulence and changes of direction. Greater wind speeds can ensure changes of air, in particular during the change of the seasons. Hence ventilation elements need to be designed in such a way that the speed of air flow can be reduced without creating whistling sounds, which may arise with narrow cracks. The summer is particularly critical as regards natural ventilation, because smaller temperature variations and frequent lack of wind mean that it is only possible to change the air by means of large openings or mechanical equipment. ⤥ **chapter 5** It is possible to boost air change rates in summer by conducting air through atriums and by using thermal effects (solar chimney).

WINDOWS
Windows are the simplest and cheapest solution as regards natural ventilation, because they are almost always present in buildings. But because they must fulfill multiple requirements (view, daylight ingress, protection against the weather, thermal comfort, low noise ingress), it is often difficult to optimize them for the purpose of changing the air. This may result in drafts and noise problems. There may be inadequate control over air distribu-

tion in the building, resulting in odor nuisance from kitchens and bathrooms. Here ventilation depends on user behavior and is therefore inconsistent. Window ventilation is suitable for locations with low wind speeds and low noise loads.

DOUBLE-SKIN EXTERNAL WALLS
In locations with a great deal of noise or wind, such as tall buildings, windows can be supplemented with a second layer. This could be in the form of a double-skin wall, a counter sash-type window or a baffle plate. This will significantly increase protection against the noise and wind, but at the same time interfere with the view to the outside. The intermediate space can be used for preheating the incoming supply air in winter and will increase the building comfort. This preheating is unnecessary in summer and a reduced number of air changes can be achieved through ventilation flaps in the outer shell. The double-skin construction provides weather protection to the sun screening elements but increases material costs. ⤥ **2**

VENTILATION FLAPS
Ventilation flaps in the facade make it possible to separate the air change function from the other functions of a window at least partially, if not entirely. This is particularly advantageous in tall buildings and in locations with strong winds. Opaque or transparent ventilation openings with substantially smaller cross-sections than windows can provide adequate air change rates, which can be adjusted further by varying the angle of the aperture. If they are of the correct size, they also offer protection against burglars and are therefore particularly suitable for nighttime ventilation. Ventilation flaps can be incorporated into a building control system and so contribute to optimizing the thermal comfort independently of the user. They can also be designed to attenuate sound and therefore present an alternative to a double-skin wall.

3,4 Examples of ventilation flaps.

5 Guide values for air changes depending on window size, number of sashes, room volume and temperature difference for side-hung or bottom-hung sashes. The graph shows the required air volume flow (m³) per hour (y-axis) based on the required air changes and the room volume (top left). The available air volume flow (m³/h) can be derived from the number of windows (top right), the type of window (side-hung or bottom-hung sash), their sizes and the temperature difference between inside and outside (bottom right).

6 Gap ventilation via window rebate vent. The volume flow is controlled automatically by mechanical means and is dependent on the wind speed. This prevents drafts.

7 Qualitative assessment of properties of different ventilation openings.

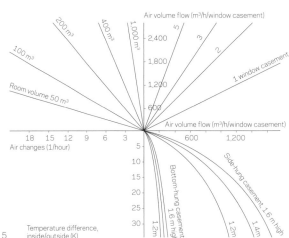

A variation on a double-skin facade is the Trombe wall. In this case the inner skin is a solid, opaque wall with ventilation flaps near the base and the top. During the day the air is heated up behind the exterior skin and can be directed by means of ventilation flaps into the interior of the building, while at night the stored heat is released into the room through the solid wall by means of a phase shift. Trombe walls are not suitable for Central Europe due to their inherently poor thermal insulation properties, but they offer possibilities for ventilating and heating buildings in Southern Europe. ⟶ 3,4

GAP VENTILATION / EXTERNAL AIR OPENINGS
In the case of gap ventilation, the necessary air change rate is achieved permanently through "planned" gaps in the building envelope. These could, for example, be air inlets in window frames. ⟶ 6 In this case the air change depends on the wind and temperature conditions. There is little risk of draughts if the ventilation openings are correctly arranged. The user is not aware of the ventilation. As these gaps cannot be adjusted, they may lead to more air changes than are required and therefore increased heat losses. There may be loss of control over the air flow distribution in the building, resulting in odor nuisance from kitchens and bathrooms.

SHAFT VENTILATION
Vertical shafts in buildings use thermal buoyancy together with air inlet elements in the facade to provide natural ventilation in buildings. This type of shaft ventilation depends on the seasons and the weather. It is also possible to ventilate internal areas by arranging the shafts in appropriate places to control the air flow distribution in the building while preventing odor nuisance.

AIR CHANGES FROM NATURAL VENTILATION
In general, it is not possible to use natural ventilation to ensure the required air changes in buildings with high occupancy rates and airtight envelopes. ⟶ chapter 5 It is possible to make a rough estimate of the air change rate that can be achieved with natural ventilation using the wind speed, the temperature difference between the interior and exterior, the window size, the number of window sashes, and the volume of the interior. ⟶ 5

6

7

	Air changes	Controllability	Sound insulation	View to the outside
Without control mechanism (existing building)	0.1–1 h⁻¹	none	-	-
Without control mechanism (new building)	0.1 h⁻¹	none	-	-
Window ventilation	1–20 h⁻¹	medium	low	good
Double-skin external wall	0.5–5 h⁻¹	low	good	low
External air vents/ gap ventilation	0.5–2 h⁻¹	good	good	-
Ventilation flap	1–3 h⁻¹	good	low	low
Ventilation flap (with control mechanism)	0.5–1 h⁻¹	good	very good	-
Ventilation flap (with sound insulation)	1–3 h⁻¹	medium	very good	-

HEAT | COOL
PLANT AND EQUIPMENT

STRATEGIES FOR DEVELOPING ENERGY CONCEPTS

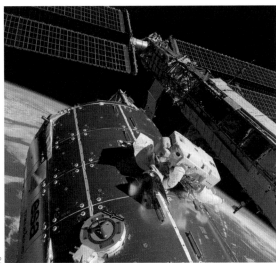

1 A sailing boat is an ideal example of the use of environmental flows of energy. The hull and sail are optimized for handling the elements and achieve the best possible performance simply by utilizing the natural energy available. However, the whole system will only function if man and machine are perfectly in tune with each other, and the external factors are also right.

2 A space station primarily needs to protect people from environmental conditions rather than exploit any natural resources. The immense technical effort is necessary in order to actively maintain the difference between the inside and outside. Both systems have a high-tech approach in common.

Effectiveness:
"the ability to get the right things done"

Efficiency (see p.93 ff.):
"the ability to do things right"

Effectiveness refers to how well a goal is achieved, i.e. the effectiveness of a concept and its resulting quality, while efficiency refers to the commercial viability of a concept, i.e. to cost and benefit, expense and return.

Intelligent concepts are characterized by the fact that the goal is achieved with few but effective means, without forcing the user into a certain mode of action. The design of plant and equipment should aim to benefit from synergy effects and to ensure that individual elements of the concept complement each other.

Not every seemingly necessary or acquired demand for comfort is really physiologically based. Some may even be detrimental to health rather than beneficial. And even reasonable needs must not automatically result in additional energy requirements if it is possible to draw upon natural energy sources instead. Therefore, an energy-efficient and sustainable building design must begin on the user and the climate. For this reason, separate chapters have been devoted to the analysis of internal and external factors, so the associated potentials and deficiencies can be identified. Chapters 5 (Building) and 6 (Plant and Equipment) use these findings and principles to derive an architectural design and a mechanical/electrical (M&E) concept.

There are three leitmotifs that can be used to develop a strategy for an energy concept. Together they will result in better efficiency in the operation of buildings:
- sufficiency (checking needs, reducing demand)
- efficiency (minimizing input and losses)
- consistency (integrating renewable energy sources)

The following process steps result from the above:
- reducing the absolute demand (substantially reducing relative demand)
- avoiding peak loads (distributing loads evenly)
- generating energy efficiently (achieving high efficiency)
- distributing energy efficiently (achieving high degree of utilization)
- reducing dependencies (making use of renewable energy)

This does not yet determine the type of energy concept or, even less so, the architectural design: solutions range from a largely self-regulating house for a thrifty user who

reacts to sudden cold spells by putting on extra clothes and accepts that he may have to sweat a little inside the building during that rare—once in a hundred years—summer, through to the highly mechanized house that provides top comfort levels at any time via automatic controls without the user having to lift a finger. Depending on the above, a different approach is required for controlling temperature, air quality, solar radiation intensity, and other factors. It follows that each concept needs a specific control strategy that includes a suitable definition of target values, control speeds, and accuracy, while taking the user into account. Depending on the concept, a slow or rapid change in the interior may be desirable; this then determines the type of construction and M&E systems.

Studies of practical applications carried out by the German government as part of the Energy Optimized Building (EnOB) research project show that the services installations of new buildings require monitoring and adjustment during the initial phase. It takes two to three years of operation before all sources of unnecessary losses are found and the full energy conservation potential is reached. Sometimes, after first occupation, buildings use twice the amount of energy compared to the amount that was calculated. The reasons are many: wrongly set operating times, incorrect target values, faulty or wrongly connected sensors, control concepts not functioning, and users acting inappropriately. Monitoring is an effective means of checking and optimizing the operation of a building. Therefore, an integrated design approach includes a building control system for the operation and monitoring of services.

LOW-TECH VERSUS HIGH-TECH?

Architects design buildings. Criteria such as energy efficiency and sustainable construction are part of the complex design process and impact on the conceptual design. Services installations can be used as a means of creating a certain design effect if they are integrated into the components of the building envelope. In any case, the solutions should always be appropriate.

Low-tech refers to a certain simplicity in the design and the use of buildings. These buildings should be constructed and operated using locally available resources. Nevertheless, low-tech construction is not synonymous with autochthonous, traditional building. Modern construction methods and materials and intelligent concepts should be integrated in the design in order to reduce the complexity and cost of M&E systems. Passive systems and self-regulating effects should be given priority. The search for ecological solutions can lead to innovative and experimental designs, but also to the rediscovery of simple building materials.

High-tech, by contrast, typically uses state-of-the-art materials and technologies that perform well, integrate new functions, and therefore conserve resources. Information systems create new user access to energy conservation; they display control concepts and facilitate the load management and automation of the building.

LOW-TECH FACADE

Air collectors integrated into a facade make use of solar energy and make it available for space heating. Intake air can be controlled by simple mechanical ventilation flaps. The integration of these components has a marked impact on the exterior design and creates a new facade architecture. Integrated thinking considers the building as a total system and activates natural potentials. Achieving the same objective with a conventional approach would involve a complicated system comprising a solar collector, buffer storage, distribution network, and radiators in order to capture solar heat and transfer it to the interior.

HIGH-TECH FACADE

The direct combination of services products and components of the building envelope often leads to a high-tech appearance. Photovoltaic cells integrated into a facade, sometimes in movable louvers, generate electricity, provide visual and solar screening, and protect openings used for night ventilation. All these functions impact on the visual design aspect.

Low-tech:
Multistory timber buildings
Straw-bale and clay buildings
(Thermal) hemp, cellulose
Air collectors, under-floor heating

High-tech:
Detachable steel and aluminum constructions
Vacuum insulation panels (VIP)
Phase change materials (PCM)
Functional glazing
Photovoltaic (PV) cells
Data bus systems
Smart meter

3 House of straw-bale construction, Unterbergern, 2006, Bauatelier Schmelz & Partner.

4 R 128, Stuttgart, 2000, Werner Sobek.

5 Institute building, Freiburg, 2006, Pfeifer Kuhn Architects.

6 Solar Decathlon House, TU Darmstadt, 2009, Energy-Efficient Building Design Group.

ENERGY SUPPLY AND ENERGY STANDARDS

The term *energy* was first introduced in its current meaning in 1852 by the Scottish physicist J. M. Rankine, and is derived from the Greek word for "work" and "activity." A distinction is made between different forms of energy, i.e. mechanical, thermal, intrinsic, electrical, magnetic, electromagnetic, chemical, and nuclear energy. Apart from nuclear energy and gravity, all energy sources on earth are derived from the sun, when one takes into account the long term conversion processes in the biosphere. Even the coal used in a coal power station is the end result of the natural process of solar irradiation on the earth. Fuels are energy sources that transport chemically bound energy, which is then converted on demand into other forms of energy. To use solar energy directly, more complex technical systems are needed in order to convert sunlight into a usable form of energy. Once harnessed, this energy is still quite difficult to store until such a time when it is needed. In the case of fossil energy sources, nature has done this work for humankind using a natural conversion process over millions of years.

FOSSIL ENERGY SOURCES
Fossil energy comes from fossil energy sources, whose energy content was transformed into concentrated form a long time ago; on a human timescale, this process is non-renewable. Fossil energy sources were created naturally through biological and physical processes over long periods of time. Crude oil, natural gas, brown coal, and anthracite are organic carbon compounds. Hence when burning these fuels with oxygen, energy is released not only in the form of heat, but depending on the composition and purity of the fossil fuel, also as other combustion products such as carbon dioxide (CO_2), nitrogen oxides (NO_x), and soot, as well as other chemical compounds.

RENEWABLE ENERGY SOURCES
Renewable energy is the opposite of fossil energy. It comes from continually regenerating processes, usually activated by the sun, and includes the use of solar radiation, wind and water power, geothermal energy, and biomass. In other words, renewable energy is derived from sustainable sources, which are inexhaustible on a human scale but are subject to frequent fluctuation. Making use of renewable energies involves diverting part of these energy flows in order to make them available to people before they are returned to the original process. Of all the renewable energy sources, wood is the source that has been used the longest. Actively burning it takes the place of its natural rotting process. In both cases, the amount of energy and carbon dioxide released is exactly the same amount that the wood has taken from the atmosphere during its growth. Thus the combustion of biomass does not alter the cycle of CO_2 assimilation and release.

RELIABILITY OF SUPPLY AND ENERGY PRICES
Germany has to import more than two thirds of the raw materials required to cover its own energy demand. The majority of heat, electricity, and fuel is derived from fossil fuels, such as mineral oil, natural gas, anthracite, and uranium, of which Germany has scarce resources. Brown coal is the only abundantly available fuel in Germany. The aspects of reliability of supply and energy prices, which are important for the economy and society at large, are determined by bilateral contracts and the development of the global energy market. Therefore, the only long-term options are to reduce energy demand locally and, at the same time, build up renewable energy facilities within the EU.

EVALUATING M&E SYSTEMS
It is not possible to achieve the objectives of an energy or climate policy without upgrading the existing building stock. According to the German Federal Statistical Office, the existing building stock accounts for almost half of all energy consumed in Germany, despite the high levels of industrial production and energy-intensive individual mobility. The previous chapters have focused on the creation of synergies during energy use and on passive strategies. Technical installations also have to be integrated into the overall design of a building. The quality of these systems can be evaluated on the proportion of energy from renewable sources and the efficiency of primary energy use.

Energy is a physical term that describes the work stored in a previously defined system or the ability of such a system to perform work.

Energy is generally measured in joules [J] or megajoules [MJ]; in the construction industry, it is more common to use watt hours [Wh] or kilowatt hours [kWh].

1 J = 1 Ws (watt second)
3.6 kJ = 1 Wh
3.6 MJ = 1 kWh
1 MJ = 0.278 kWh
(1 h = 3,600 s)

As early as 1963, in his book "Operating Manual for Spaceship Earth", Buckminster Fuller referred to the limited supply of fossil fuels powering the starter motor for human civilization, which should be replaced by sustainable (renewable) energy sources as soon as possible.

1 The entire energy available from various renewable fuels is illustrated in the form of large cubes. The small cubes shown within these represent the proportion of energy which, in theory, is currently technically and economically accessible to us. Of this accessible energy, we only actually use a small fraction. This is compared with the actual world energy consumption in the year 2000 and the projected world energy demand for the years 2020 and 2050. This clearly shows the enormous potential for satisfying the world's increasing hunger for energy from renewable energy sources. In view of this, it appears all the more urgent to convert the theoretical potential into practical exploitation.

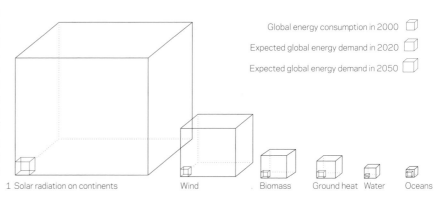

Global energy consumption in 2000
Expected global energy demand in 2020
Expected global energy demand in 2050

1 Solar radiation on continents   Wind   Biomass   Ground heat   Water   Oceans

2

3

2 Solar Decathlon house 2007.

3 Solar Decathlon house 2009.

Another form of assessment, which is not referred to in EnEV, uses CO_2 equivalents, a method that measures emissions of gases with climatic effect per unit of final energy used. While the study of primary energy and CO_2 focuses on global climate effects, there are other emission factors that may have an effect on smaller regions. An example is the emission of fine dust from wood stoves.

Until 2007, for example, the German Credit Institute for Reconstruction (KfW) linked its CO_2 building retrofit program (hence the name) with the stipulation that CO_2 emissions per square meter and year should be reduced by 40 kg as a result of the retrofit works.

For different methods of calculation using net, final, and primary energy (see p. 36 f.).

SYSTEM QUALITY EXPRESSED QUANTITATIVELY

Efficiency describes the relationship between the achieved output or performance and the input required to achieve it. If these are measured in terms of energy, this describes a system's degree of efficiency. The reciprocal value is called the *ep-value* (HVAC coefficient). Reducing losses in generation, storage, and distribution increases efficiency and reduces system running costs. The degree of efficiency must not be confused with the degree of utilization; the former is determined for certain fixed parameters and is therefore constant for these operating conditions, whereas the latter relates to a period of time, usually a year, enabling the evaluation of different operating conditions. Losses during standstill and unused excesses decrease the degree of utilization.

PRIMARY ENERGY APPROACH

In a closed system, energy cannot be destroyed (law of conservation of energy). But energy may be present in a form that cannot be used by people, meaning it is not possible to physically generate or consume, lose or conserve. The only options are to minimize demand or make more efficient use of the energy. Thus the concept of *energy generation* is, strictly speaking, not exactly correct; it refers to a transformation of primary energy from one form that is not usable by people—or cannot be easily used—into another, more usable energy form. It is important to distinguish these different forms as they belong to different calculation methods. The term *primary energy* refers to the energy available in the naturally occurring energy forms or sources. The amount of energy made available to the consumer after further losses from conversion and transmission is called the *final energy*. The different energy types are associated as follows:
The different types of energy are associated as follows:
- net energy: for domestic hot water and space heating
- final energy: heating oil, wood pellets, mains electricity
- primary energy: crude oil, crude gas, pit coal, unprocessed timber, raw uranium.

While the Thermal Insulation Regulations (WSVO) govern the requirement for space heating, i.e. a form of net energy, the Energy Conservation Regulations (EnEV) focus on primary energy: the primary energy required for supplying a building with heat and electricity is evaluated. This approach is designed to ensure that the entire chain of losses, from the energy source through to the final use, is taken into consideration. As a consequence, many building energy standards have been aligned to this approach. The German standard for new construction is defined by the EnEV and the REHA (Renewable Energies Heat Act). Various building standards exist for energy concepts that exceed the above.

NEW CONSTRUCTION AND "EFFICIENCY HOUSES"

Subsidies from federal and state governments are tied to the reference building of the currently applicable EnEV. Examples are the KfW Effizienzhaus levels and model projects by the German Energy Agency (dena), such as the "low-energy house" program for existing buildings. A percentage reduction in annual primary energy use and improved exterior thermal insulation are required.

PASSIVE HOUSES

Due to its highly insulated building envelope, a house built to Passivhaus standards does not need a conventional heating system. The heat requirement is largely covered by thermal gains from solar irradiation, waste heat from people, and technical equipment. The remaining space heating requirement of 15 kWh/m²a can be satisfied by controlled ventilation with heated intake air. Additionally, no more than 40 kWh/m²a of primary energy may used to supply the heat. Triple glazing and thermal insulation that may exceed 30 cm in thickness are required.

NET ZERO ENERGY HOUSES

A zero energy house has an annual balance of null: peak loads are covered by the mains network an excess generated electricity is fed back to the network. This building concept uses the public utility network to balance seasonal energy discrepancies. It combines decentralized energy generation with centralized load management. The calculations are usually made on the basis of primary energy. A zero annual balance is only possible in theory, as the building will overrun or fall short of the zero balance depending on the weather in any one year.

PLUS ENERGY AND SELF-SUFFICIENT HOUSES

The plus energy house is supposed to achieve a surplus in the annual balance, irrespective of the weather. These houses are connected to the public utilities, whereas self-sufficient houses are not. The latter cover their annual requirement for electricity and heat with buffer storage, independently of the network.

NET ZERO CO_2 EMISSION HOUSES

A building without CO_2 emissions requires the use of renewable energy sources and has objectives similar to the zero primary energy house. Benefits from photovoltaics and cogeneration (CHP) can be entered on the credit side in the calculation.

NET ZERO ENERGY COST HOUSES

Payments received for feeding electricity into the network reduce the user's already low energy cost to zero, but increased investment costs must be financed.

USE-SPECIFIC ENERGY REQUIREMENT

Cellar (ancillary room)
Stairwell (circulation area)
Apartments (multi-family dwelling)

kWh/m²a   0   50   100   150

Sample calculation for multi-family dwelling
Cellar: 1.0 kWh/m²a x 200 m² = 200 kWh/a
Stairwell: 1.5 kWh/m²a x 200 m² = 300 kWh/a
Apartments: 70 kWh/m²a x 1,600 m² = 112,000 kWh/a

200 m²
Cellar
200 kWh/a

200 m²
Stairwell
300 kWh/a

1600 m²
Apartments
112,000 kWh/a

Sanitary facilities/WC
Staff room (group office)
Auditorium (theater foyer)
Assembly room (circulation area)
Gymnasium 2-court
Classroom

kWh/m²a   0   50   100   150

150 m²
Sanitary
facilities
4,350 kWh/a

300 m²
Staff room
21,900 kWh/a

300 m²
Auditorium
6,000 kWh/a

600 m²
Assembly room
6,600 kWh/a

1200 m²
Gymnasium 2-court
91,200 kWh/a

1250 m²
Classroom
54,375 kWh/a

RESIDENTIAL
The analysis focuses on the specific and absolute net energy requirement for the various services, i.e. domestic hot water, space heating, cooling, artificial lighting, mechanical ventilation, and household equipment. EnEV does not contain any provisions regarding the energy requirement of artificial lighting and household equipment; nevertheless these elements are shown here for greater clarity. The resulting distribution is typical for energy optimized buildings that make use of synergies and thereby exceed the EnEV 2009 standard for new buildings. This does not achieve the Passivhaus standard, which can be considered as the optimum reduction of demand that is still commercially viable.
The amount of energy required in residential buildings is determined by the demand for space heating. Mechanical cooling of the building should be avoided in favor of passive means. Solar thermal installations are typically used to contribute to the preparation of domestic hot water and space heating. There is practically no energy demand in the stairwell and cellar zones, as these zones are not heated and are assumed to be naturally ventilated. Where part of the cellar is used as a laundry room, some of the electricity used in the rooms above would now be used in the cellar so that in total there is no increase in consumption.

SCHOOLS
Where optimum comfort levels must be achieved throughout the year, classrooms and offices in school buildings have to be provided with heating and cooling. Owing to the dense occupation of these rooms, mechanical intake air is required. This results in a considerable demand for energy for ventilation. At the same time, and because volumetric flows are high, the ventilation system can be used to provide heating/cooling to the rooms. Therefore the analysis of a services concept for school buildings should start with the ventilation system. Ancillary rooms and areas that are only used temporarily can be integrated in the system where possible.
The gymnasium needs to be considered separately. It is often a separate building and is also frequently and for prolonged periods used by third parties. Owing to the large volume and type of use, this building has an increased demand for heat, particularly for the provision of domestic hot water for the showers, which may also exist over the holiday/vacation periods.

OFFICES

More than half the net energy demand in the main areas of office and administration buildings is in the form of electricity. It is therefore logical that the EnEV regulations stipulate that artificial lighting and ventilation are included in the energy calculations for non-residential buildings. User-specific equipment (computers, monitors, printers, copiers, desktop lights, etc.) is not included in the calculation methods prescribed in DIN V 18599 as it is considered part of the movable furniture. However, as the diagram illustrates, these items of equipment already account for more than a quarter of the energy demand in terms of net energy and a third of the primary energy demand. It is therefore imperative to select energy-efficient equipment in order to reduce the demand for electrical energy as much as possible. In any case, this equipment needs to be considered as part of the overall energy concept as the heat it gives off adds significantly to the internal heat output. While the demand for heating and cooling is high, the required air changes are relatively low, which would speak in favor of using building components for thermal storage and transfer purposes.

INDUSTRIAL BUILDINGS

Similar to the office equipment used in administration buildings, it is the machinery for the production and processing of goods that determines the energy requirement in industrial buildings. This has to be calculated for each specific project and use. With the help of heat exchangers and heat pumps, it is possible to utilize the waste heat from equipment as an energy source for the building, to heat the domestic hot water, for example. This can be done indirectly through the heated exhaust air, but should preferably be done directly by connecting the supply pipes to the cooling water circuit of the machinery. The secondary power requirement for operating the services in industrial buildings is usually small, as they are naturally ventilated and heated by ceiling radiating panels.

- Domestic hot water
- Heating
- Cooling
- Lighting
- Ventilation
- Work equipment

TRANSFER SYSTEMS FOR VENTILATING, HEATING, AND COOLING

Transfer systems may be classified according to the four different forms of construction discussed below. The simplest system for heating a room involves heating elements, but it is also possible to utilize building components for the purpose of heating/cooling. Where ventilation systems have been installed, these can also be used for heating/cooling interiors. In commercial buildings it is common to find heating/cooling elements in ceilings.

CONVECTION AND RADIATION

Convection transfers heat through air flows, which carry the thermal energy. Thermal radiation transfers thermal energy from one place to another without involving any mass; it is not dependent on any heat carrier medium and is more efficient. Thermal radiation from surrounding surfaces is perceived by people as more comfortable than heating the room air. Unpleasant drafts only occur in convection systems with a high temperature differential between the heat source and the room air. Radiators should be installed on external walls to avoid asymmetric radiation patterns. ➔ p.43 Individually controlled radiators should be fitted with a return valve and a flow valve with a display of the setting in order to facilitate maintenance. New radiators should be designed for the heat generation plant used but should not exceed 60/40 °C for flow and return. Particularly important is a low return temperature so that the condensing effect of boilers can be utilized and, in the case of district heating, reduced volume flows can be achieved.

CONVECTORS

Convectors are heat exchangers that transmit 85% of their heat to the room by convection. Fins welded around the heating pipes provide a large surface area for warming the surrounding air that flows past. The simplest convector form is a heating pipe with helical fins around it. Underfloor convectors are integrated into the floor construction for space heating with no visible heaters. This is useful for supplying heating near floor-to-ceiling glazing along exterior walls, and convectors can also be installed at any height along the facade. Fan-assisted convectors can improve the performance where there is only a slight difference between the convector temperature and the room air. This is also advantageous for heating rooms with great depth; furthermore, these units can also be used for cooling purposes. Fan convectors can be operated with recirculating air or with fresh intake air. Negative aspects are the relatively high purchase price and the noise from the fan. Where mechanical ventilation systems are used, convectors should not be installed, as these could interfere with the ventilation air flow.

RADIATORS

Radiators are hot water heating systems that emit a considerable proportion of their heat by radiation (max. 40%). Units of the same output are larger than convectors, but have a relatively small surface area for heat exhange. They are made of cast iron or steel, through which the hot water flows. Compared to convectors, they cause less dust circulation. As their water content is relatively small, radiators can respond rapidly to individual heat requirements. For radiators that heat the air well and also radiate heat, convector sheets are combined with flat plate radiators. In double and triple style radiators, the flow of hot water runs first through the front panel before being conducted through the second or third panels. During normal operation, the output from the front plate is fully sufficient and the second or third plates will hardly be heated. Only when there is an increased demand for heat output will these plates contribute to heating the room rapidly with convection. The advantage is that the time taken to heat up the room is reduced by up to 25%.

NIGHT STORAGE HEATERS

Night storage heaters are electrically operated devices in which heat storage elements are heated during "off-peak" periods using electricity at reduced cost. The heat is usually stored in heavy refractory blocks of magnesite, which are heated overnight to temperatures of up to 650 °C. Preheating the heater in this way is subject to some margin of error, as the heat requirement of the following day has to be estimated using an outside temperature sensor. During the day the heater gives off heat to the room with the help of a fan. Compared to other forms of heating, the overall degree of utilization is poor, so the use of such systems is not recommended.

BASEBOARD HEATING

When the flow temperature is well adjusted, a thin veil of warm air rises slowly on the surface of the wall above the baseboard heating element (Coanda effect). The heat content of the rising air is transferred to the wall and from there to the room in the form of long-wave heat radiation. This is the difference between baseboard heating and underfloor convectors. After the veil of air has cooled down to room temperature, it horizontally moves into the room and then slowly sinks, taking dust particles with it to the floor. This achieves an even heating of the entire room as well as a healthy room climate. The system is particularly suitable for existing buildings where it is not possible or desirable to have underfloor heating. At a flow temperature of 45 °C these systems produce a heat output of about 140 W/m. The heating up phase is short. The system is also suitable for large glazed areas and can be fitted concealed behind furniture.

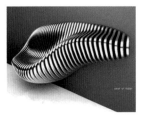

Radiator as design element.

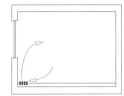

Convector.

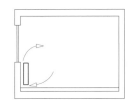

Underfloor convector.

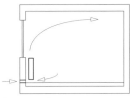

Fan-assisted convector.

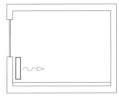

Radiator.

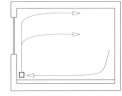

Baseboard heating.

High output ceiling cooling.

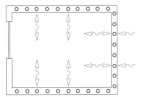

Building component activation.

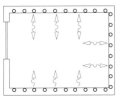

Radiant wall heating and cooling.

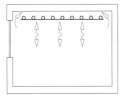

Ceiling panels.

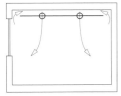

Chilled beams.

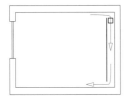

Peripheral cooling.

Heating and cooling systems involving large surfaces create good levels of comfort due to the large heat exchange area. In heating mode the proportion of radiation is nearly 100% and in cooling mode approx. 60%. The remaining convective part is mostly generated by internal heat sources which create a rising flow of warm air. As these systems work with low-flow temperatures, they lend themselves better to the use of alternative energy sources. Due to the higher inertia of these systems, rapid heating up is not possible; on the other hand, disagreeable temperature fluctuations in the room are avoided.

BUILDING COMPONENT ACTIVATION / THERMO-ACTIVE CEILING

Building component activation (also called concrete core activation) refers to systems in which pipes are integrated in solid walls and slabs during construction. In this way the entire component is used for heat transfer and storage. The activated mass levels-out peak heating and cooling loads. In practice, separate heating and cooling loops are installed at different facade locations to compensate for overheating or cooling in certain zones. The average flow temperature in the loops is 23 °C throughout the year. A temperature difference of 3–5 °C provides cooling in summer and heating in winter. Since it is not possible to install suspended ceilings or raised floors with such systems, acoustic treatment and flexibility of use of the space are impacted. Nevertheless, this system is frequently used in office and administration buildings. To increase the control speed, the basic system can be supplemented with cooling panels or conditioned supply air.

HEATING AND COOLING CEILINGS

Heating and cooling ceilings are closed systems with pipes installed near the ceiling surface, embedded in the plaster. Insulation is placed between the active surface layer and the slab above, which distinguishes these systems from those using building component activation. Heat is transmitted or absorbed (cooling) only from the lower face of the ceiling. Thus the specific output is lower with freely suspended ceiling panels. Normally about 75% of the available ceiling area is used for the purpose of ceiling heating/cooling. For efficient placement, ceilings are best suited to activation for cooling, whereas for heating, walls and floors are more suitable locations.

FLOOR AND WALL HEATING

In underfloor heating systems, the heating pipes are embedded in the screed (wet screed) or under the screed board (dry screed), in which case they are used in conjunction with heat spreading plates. Insulated plastic panels with molded nodes make installation easier. With wall heating systems, the pipes are usually mounted on thin pipe mesh and covered with plaster. Thick clay plaster has the advantage that it can regulate the relative humidity as well as perform the heating function. However, it must be borne in mind that the thickness of the plaster impacts on the inertia of the system. With underfloor heating it is important to choose the right floor finish to allow the transfer of heat. Tiles and stone materials have good storage capacity and transfer heat evenly. Wooden floors are also suitable, but carpet will reduce the output. Both systems can be used for heating and cooling but preference should be given to the heating function, as underfloor cooling could easily lead to a sense of discomfort. With walls it is only the radiation that is perceived, which induces a sense of comfort. Therefore, wall heating can be used for cooling and heating without loss of comfort.

CEILING SAILS, PANELS, AND STRIPS

The heating and cooling elements are suspended from the ceiling in various forms of sails, panels, and strips. They contain metal plates connected to pipes conducting a heating/cooling medium. They achieve increased outputs as they are freely suspended and surrounded by air (without insulation above). The ceiling of the room is also activated by radiation and helps to heat/cool the surrounding air. With a significantly smaller area (about 30–50% of the floor area), these systems achieve the same heating/cooling output as an entire ceiling, and the investment cost is proportionately lower.

CHILLED BEAMS

Chilled beams can be suspended freely or installed flush with a perforated suspended ceiling. Warm air rises in the room, cools down at the chilled beam, and sinks back down to the occupied area due to its greater density. Active chilled beams are a fan-assisted variant.

PERIPHERAL COOLING

In peripheral cooling the chilled beam principle is applied to the wall. A convector with cold water feed pipe is installed near the ceiling, concealed behind a panel. This will cool the warm air that has risen to the ceiling, causing it to convect downwards. Near the floor the cool air can flow from behind the panel and return to the room.

DIRECT ELECTRIC HEATING

Direct electric heating may be installed as underfloor and wall heating and should be placed as close as possible to the surface in order to allow the surfaces to heat up rapidly. These systems are usually not economical to run as the heating-up time falls into a high tariff period. Therefore they should only be used for small rooms with temporary heating requirements, such as bathrooms.

In industrial buildings the transfer of heat poses a particular problem due to the geometry and the use of such spaces: panel heating systems are often not suitable for structural reasons or due to lack of flexibility; in addition their output is usually insufficient. The heated room air accumulates in heat buffers underneath the ceiling, far from the users, and large heat losses occur when these buildings are opened to the outside. In this situation people will not feel comfortable, even when performing physically intensive work.

A high temperature radiation heating system presents a solution. Infrared radiation does not require a carrier medium for transporting its energy. Heat from the unit reaches people directly, without deviation and almost without any loss, and also does not cause any draft. The heat gain from radiation compensates for heat losses to cold room air, achieving an acceptable thermal balance. The room air is not heated directly, but warms slightly on contact with walls and floor. Compared to other heating systems, the temperature distribution is uniform both vertically and horizontally. Typical applications are therefore industrial and commercial buildings, sports halls, and churches.

The installation of high temperature radiation heating systems in large buildings has the following advantages:
- compensation for low air temperature through radiated heat in buildings with poor insulation or that are frequently opened to the outside; suitable also in situations where raw materials or machines (e.g. trains) have to be allowed to cool down before being worked on;
- higher wall and floor temperatures compared to conventional heating systems, better thermal comfort with air temperatures lower by 2–3 °C, no dust convection;
- lower ceiling and air temperatures, no uncomfortable drafts, overall reduced space heating requirement;
- no moving parts therefore practically maintenance free, low operating costs;
- installation on the ceiling means that floor and wall surfaces can be fully utilized.

LUMINOUS INFRARED GAS HEATERS
Luminous infrared gas heaters are a version of infrared heating used in spaces higher than 6 meters. The infrared radiation is generated by open combustion of a gas/air mixture. The exhaust fumes must be indirectly disposed via the room air, requiring that fresh air is supplied at a rate of 10 m³/h per kW. It is therefore important to ensure adequate natural ventilation, which, however, is usually already needed for the normal activities that take place in these types of premises.

RADIANT TUBE INFRARED GAS HEATERS
Radiant tube infrared gas heaters usually consist of U-shaped tubes, through which combustion gas is fed. A reflector is attached above the entire length of tubing, which is a closed system, and directs the heat to the areas in need of heating. The average surface temperature is 250–500 °C and is almost the same throughout the length of the tube. These systems are usually used in halls with a height of 4–8 meters. Where several burners are connected to a shared supply line, the exhaust gases can be removed through a combined discharge system. If a heat exchanger is included in the system, it is possible to recover both sensible and, if any, latent heat, which can be used for heating adjacent offices. In this way it is possible to heat office areas of up to 20% of the hall area. It is also possible to heat hot water for showers.

RADIANT STEAM HEATERS AND CEILING RADIATION
These systems function similarly but at lower temperatures; the steam can be produced by heating the water with a range of boilers, not only gas burners.

POSITIONING THE RADIANT HEATERS
A distinction must be made between full space heating of the hall and partial or workplace heating. The manufacturers' specified minimum suspension heights must be observed. Full space heating allows greater flexibility and comfort but requires greater energy input as the area being heated is larger. For each 1 °C increase in temperature, approx. 40 W/m² is required for workplace heating, whereas only 14 W/m² is needed for full space heating.

Infrared radiation devices must be laid out to ensure that they cover the required area completely. The units are positioned such that their radiation cones overlap (by 90°) at the edges. Owing to the special physics involved in radiation heaters, the heat output often has to be oversized to meet standard requirements. The rating of these units can be calculated and optimized in accordance with the DVGW G 638 worksheets. These cover the basic physiological requirements, options for the discharge of exhaust gas from luminous infrared gas heaters, and installation instructions regarding the height and distance of units.

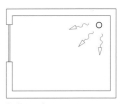

Ceiling radiator.

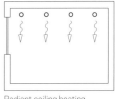

Radiant ceiling heating.

The diagram shows an example of the correlation between radiation intensity, air temperature, and the resulting perceived room temperature. A k-factor of 0.033 is typical for ceiling-mounted full space heating systems.

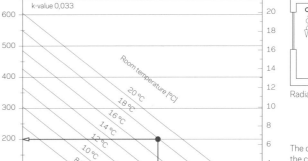

Radiation intensity [W/m²]   Radiation temperature [°C]

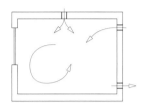

Mixed mode ventilation.

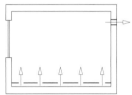

Trickle ventilation.

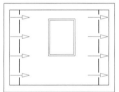

Displacement ventilation.

Natural night ventilation.

Exhaust air with decentralized intake air through window frames.

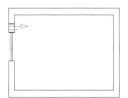

Central intake and exhaust air through roller shutter boxes with heat recovery.

The main task of mechanical ventilation systems is to renew stale and polluted room air. Where natural convection is not sufficient, fans are used to distribute the air. In contrast to window ventilation, this ensures a constant comfortable room climate.

MIXED MODE VENTILATION / INJECTION VENTILATION
In mixed mode ventilation, the supply air is forced through air inlets in the ceiling or wall at slightly increased speed. This intake air mixes with the room air and people in the room obtain an almost homogenous mixture of fresh and stale air. Where the ceiling outlets are flush with the ceiling or wall outlets are located no more than 30 cm below the ceiling, this air injection spreads horizontally across a wide area. This is particularly important where cooling is concerned in order to avoid any draft reaching people. These outlets are not usually suitable for heating high rooms since the warm air would tend to remain at a high level under the ceiling. However, in rooms less than 3.5 m high full space heating will be achieved after some time. Air is extracted through air outlets near the floor or by overflowing into adjoining air extraction areas such as WCs.

TRICKLE VENTILATION
Trickle ventilation is a supply air system with extremely low turbulence. Low air speed reduces the induction effect, mixing of stale room air with intake air is minimized. The intake air should enter the room via inlet slots or diffusers at a slightly lower temperature (2–4 °C lower than the room air temperature). This will create a cold air "pond" of fresh air. Through the air rising from heat sources (users) it is possible to discharge most pollutants. This makes the system attractive for offices as well as industrial premises with pollutant emissions. For the design of this type of system it is necessary to know the heat sources present, as the resulting convection affects the air flow. For comfort reasons, the temperature of the intake air, which also defines the limit of the system's performance, must not be too low.

DISPLACEMENT VENTILATION
In contrast to the ventilation systems above, supply ventilation in the form of large area displacement ventilation is only used in special rooms such as laboratories or operation theaters. In this system, the supply air is introduced evenly across the entire wall or ceiling surface, providing the cooling before being extracted on the opposite side. The outgoing air flow must correspond exactly to the intake air flow and is usually extracted through a perforated raised floor or a perforated wall.

FREE NIGHT VENTILATION
Utilizing the lower nighttime temperatures can help to compensate for heat gain during the day. For effective cooling, the outside temperature should be below 21 °C every night for at least 5 hours and the air change rate in the building should be least 2–4 h-1. The cooling effect is achieved through open surfaces of storage mass in the rooms. It is therefore advantageous if ceilings and walls are in solid concrete construction, and suspended ceilings and raised floors should be omitted where possible. A prerequisite for free night ventilation is good outside air quality, sufficient air flow through the building, and openings protected against intruders and noise from outside. The cooling effect can be improved using convective or wind-induced air flow, which requires vertical air spaces or openings in opposite facades. It is possible to use automate the night cooling with the help of indoor and outdoor sensors to control the process.

MECHANICAL NIGHT VENTILATION
Where special openings in the facade are not possible (for security or other reasons), night ventilation can be provided by fully mechanical means using air intake and extraction. In this case, the additional energy input needed must be compared with the energy required for active cooling. Where natural ventilation is only restricted by the lack of convection, an air extraction system will suffice.

RESIDENTIAL VENTILATION SYSTEMS
Such systems are typical of houses built to the Passivhaus standard. A central ventilation unit installed in the cellar or the attic feeds air to most rooms via supply air ducts and receives exhaust air from kitchens and bathrooms. Fresh air with low dust and odor levels should be obtained via the roof or from the yard, at a sufficient distance from the exhaust air outlets. The ventilation unit comprises two fans (for intake and exhaust air), filters for dust and pollen, a heat exchanger for recovering heat and sometimes humidity, and a regulating system with thermostat controller in the dwelling. The air ducts are fitted with sound insulation to suppress noise from the fans and sound transmission between rooms.

DECENTRALISED VENTILATION ABOVE WINDOWS
As an alternative to the more complex central ventilation systems, it is possible to integrate decentralized air intake and discharge systems with heat recovery in roller shutter boxes. This makes it easier to carry out upgrades when retrofitting a house. The system has an advantage over other decentralized solutions, in that it is not visible in the facade. Likewise, its position above the window means that the interior is not restricted in any way, and no additional openings are required in the wall.

PERFORMANCE OF DIFFERENT TRANSFER SYSTEMS AND DEGREE OF COMFORT ACHIEVED

The comfort level in a room is largely affected by the building envelope properties. This is particularly the case in summer, when a number of functions must be accommodated (solar screening, glare protection, provision of artificial and daylighting, unobstructed view, ventilation) and no clear-cut optimization is possible, but compromises have to be found. Comfort levels in summer are evaluated via the hours of excess temperature in a room (over 26 or 28 °C respectively) and the maximum room temperature.

The illustration on the right shows simulations of the level of comfort resulting from the facade design (proportion of window area f, total energy transmittance g, solar protection factor Fc) and the type of system for transferring air, heating, and cooling to the room. With a low proportion of window area and good solar protection, it is possible to achieve good levels of comfort with purely passive cooling methods. By contrast, where maximum comfort is required as well as a high degree of freedom in the facade design, it is necessary to utilize building components for cooling (building component activation).

uncomfortable

acceptable

comfortable

very comfortable

The table on the left shows most of the thermal transfer systems described above. The typical flow and return temperatures of the system are stated (for radiation devices: reflector temperatures). The ideal temperature range of radiators, wall, floor, and ceiling heating systems and of intake air is limited by the user's sense of comfort. Depending on other parameters (such as volumetric flow, etc.), these systems have a typical maximum heating and cooling output relative to the floor area of the room.

1 Radiator as room divider.

2 Ceiling cooling element.

3 Heating and cooling capacity of different systems.

System	System temperature [°C] cooling / heating		Ideal temperature range	Output [W/m²] cooling / heating		Comments
Radiator	-	up to 70/50 °C	35-55 °C	not common	> 100 W/m²	economical, simple controls
Fan assisted convectors	up to 12/16 °C	up to 70/50 °C	16-55 °C	> 50 W/m²	> 100 W/m²	considerable drafts and high electricity consumption when cooling output is high
Underfloor heating/cooling	up to 12/16 °C	up to 40/35 °C	19-28 °C	< 30 W/m²	< 60 W/m²	higher heating/cooling output reduces comfort level
Ceiling heating/cooling	up to 12/16 °C	up to 40/35 °C	16-30 °C	< 80 W/m²	< 40 W/m²	very efficient and good comfort levels through low system temperatures
Wall heating/cooling	up to 12/16 °C	up to 40/35 °C	18-34 °C	< 50 W/m²	< 80 W/m²	very efficient and good comfort levels through low system temperatures
Concrete core activation	up to 18/24 °C	up to 30/28 °C	18-24 °C	< 50 W/m²	< 25 W/m²	advantages similar to ceiling heating, individual room control very limited
Circulation air	up to 12/16 °C	up to 70/50 °C	16-55 °C	< 100 W/m²	< 100 W/m²	considerable drafts and high electricity consumption when output is high, not comfortable
Combined ventilation	up to 12/16 °C	up to 40/35 °C	16-40 °C	< 60 W/m²	< 30 W/m²	low additional cost of 5.00 €/m² when mechanical ventilation is required anyway
Trickle ventilation	up to 12/16 °C	-	3 °C below room air temp.	< 25 W/m²	not possible	low draft, high level of comfort, very low output
Displacement ventilation	up to 12/16 °C	-	up to 8 °C below room air temp.	< 50 W/m²	not possible	low draft, high level of comfort, low output
Adiabatic cooling, direct/indirect	-	-	reduction by 2-4 °C / 3-5 °C	< 40 W/m²	not possible	limited as it increases relative humidity (direct in intake air, indirect in exhaust air)
Night air cooling	-	-	external temp. below 21 °C			draughts are acceptable when no person present, natural (cross) ventilation is preferable
Luminous radiant gas heater	-	up to 950 °C	-	not possible	> 100 W/m²	radiation compensates for low air temperature, temporary operation is possible
Radiant tube infrared gas heater	-	250-500 °C	-	not possible	> 100 W/m²	combined exhaust gas discharge improves room air and makes it possible to install heat recovery
Radiant steam heater	-	100-150 °C	-	not possible	> 100 W/m²	no gas combustion in the room, can be connected to a central heat generating source
Radiant ceiling panel (water based)	-	70-90 °C	-	not common	> 100 W/m²	low temperature makes it possible to use process waste heat where available

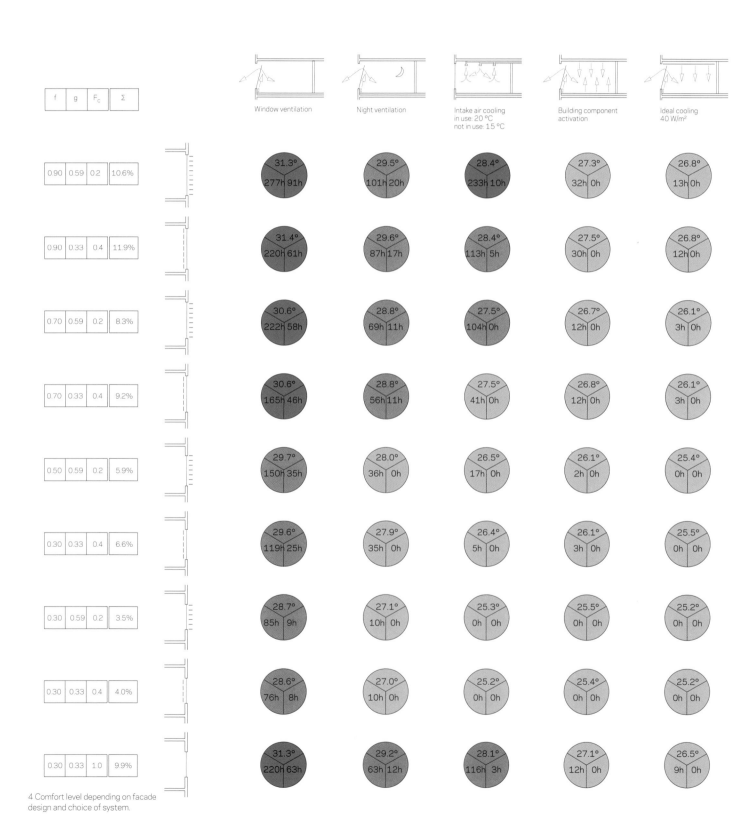

f	g	F_c	Σ

Window ventilation

Night ventilation

Intake air cooling
in use: 20 °C
not in use: 15 °C

Building component activation

Ideal cooling
40 W/m²

f	g	F_c	Σ
0.90	0.59	0.2	10.6%
0.90	0.33	0.4	11.9%
0.70	0.59	0.2	8.3%
0.70	0.33	0.4	9.2%
0.50	0.59	0.2	5.9%
0.30	0.33	0.4	6.6%
0.30	0.59	0.2	3.5%
0.30	0.33	0.4	4.0%
0.30	0.33	1.0	9.9%

Row 1: 31.3° 277h 91h | 29.5° 101h 20h | 28.4° 233h 10h | 27.3° 32h 0h | 26.8° 13h 0h
Row 2: 31.4° 220h 61h | 29.6° 87h 17h | 28.4° 113h 5h | 27.5° 30h 0h | 26.8° 12h 0h
Row 3: 30.6° 222h 58h | 28.8° 69h 11h | 27.5° 104h 0h | 26.7° 12h 0h | 26.1° 3h 0h
Row 4: 30.6° 165h 46h | 28.8° 56h 11h | 27.5° 41h 0h | 26.8° 12h 0h | 26.1° 3h 0h
Row 5: 29.7° 150h 35h | 28.0° 36h 0h | 26.5° 17h 0h | 26.1° 2h 0h | 25.4° 0h 0h
Row 6: 29.6° 119h 25h | 27.9° 35h 0h | 26.4° 5h 0h | 26.1° 3h 0h | 25.5° 0h 0h
Row 7: 28.7° 85h 9h | 27.1° 10h 0h | 25.3° 0h 0h | 25.5° 0h 0h | 25.2° 0h 0h
Row 8: 28.6° 76h 8h | 27.0° 10h 0h | 25.2° 0h 0h | 25.4° 0h 0h | 25.2° 0h 0h
Row 9: 31.3° 220h 63h | 29.2° 63h 12h | 28.1° 116h 3h | 27.1° 12h 0h | 26.5° 9h 0h

4 Comfort level depending on facade design and choice of system.

AIR- AND WATER-BASED DISTRIBUTION NETWORKS

A number of technical networks must be installed to service a building. For their efficiency, it is important that these systems are designed to suit the building's structure. With respect to the supply and discharge of heat, a distinction is made between air- and water-based systems.

Hot water heating uses a closed water circuit as carrier medium. Modern systems use separate pipes for flow and return (two-pipe, or flow and return system). The distribution network is subdivided into three parts. The distribution pipe runs horizontally underneath cellar ceilings or within the first floor construction. It connects between the heat generating source and the riser, which conducts the hot water to the upper floors. In existing buildings, risers are usually placed within or on external walls, but in new buildings they may also run within partition walls. Finally there are link or spur pipes, which connect the network to individual heat transfer elements. These lines are usually installed within the floor, and less often in horizontal slots in the walls (reduced structural cross section). In larger buildings, the vertical risers are centrally combined in shafts feeding all the floors. The position of these shafts often imposes limitations if a building is modified and the layout is to be changed. The position and size of these shafts also affects the design of other installations. Empty conduit can help to connect solar thermal installations at a later date. Depending on the use requirements and internal factors (flow temperature and operating times), independently controllable heating circuits should be installed, down to individual room control but at least separated according to different facade orientations (NE, SW).

Heating pipe is often made of copper, which comes in rolls of up to 50 m length. This makes it possible to avoid pipe joints in inaccessible places. Copper pipe can be bent and cut to size manually, thus it also has a time advantage compared to steel pipe. The latter is less expensive, but due to the laborious jointing (welding, compression fittings) and risk (albeit low) of rusting from the inside, has only a small market share. Modern oxygen-proof plastic composite pipes are also making inroads. They can perform safely at up to 90 °C and with operating pressures of 6–10 bar. The pipe-in-pipe system is popular, in which a corrugated outer pipe protects the inner pipe from mechanical damage and absorbs any temperature-induced expansion. Damaged internal pipes can be replaced even after long periods of time, but a disadvantage of the pipe-in-pipe system is its larger diameter. Potable water lines are usually stainless steel composite pipes, sometimes copper or plastic.
Pipes must be insulated; requirements are listed e.g. in Appendix 5 of the EnEV regulations. This increases the overall diameter threefold, since the minimum thickness of the insulation (thermal conductivity 0.035 W/mK) should equal that of the pipe's diameter. In buildings with very low space heating requirements but a constant demand for domestic hot water (e.g. a building to Passivhaus standards), it is recommended to exceed the EnEV requirements by fitting insulation of up to twice the diameter.

To provide domestic hot water without any waiting time at the faucet, circulation pipes are installed. In this system, unused water is returned to be reheated at the heat source; the higher water temperature reduces the risk of Legionella contamination. Circulation pipe systems consume more energy and are more expensive to run, but save water and provide better comfort. They can be found as standard installations in multifamily dwellings and non-residential buildings. Where the intention is to reduce the amount of pipe installation and heat loss, it is possible to omit hot water at hand wash basins. This is particularly applicable for janitor's rooms. An alternative is to use a fresh water system: in this system a cold water supply is fitted with a heat exchanger connected to a heating system supply line. In this case hot water is not produced until it is needed at the faucet.

Electrical circulation pumps transport the heat carrier medium or the drinking water in the various circuits. The large number of pumps (heating and cooling circuits, solar circuits, circulation systems, etc.) leads to a considerable demand for secondary electricity. In unfavorably designed systems, up to 15% of the primary energy is used for pumping alone. In single family houses, electricity charges of over EUR 150 p.a. are not uncommon, so it can be worthwhile to replace older pumps. It is often possible to reduce the required power input to a fraction, meaning the investment pays for itself within a few years. In the future it will also be possible to install decentralized pump systems (directly at the radiators). Studies of these systems carried out by the Fraunhofer Institute have revealed significant savings potential for heating energy and secondary electricity of 30–50% respectively.

Despite the increased efficiency of pumps, all loops should be hydraulically balanced upon commissioning and after maintenance work. Hydraulic balancing is a process in which, at a set flow temperature specified for the heating system, all heat emitting elements are supplied with exactly the right amount of heat necessary to achieve the desired room temperature in the individual rooms.

1 Underfloor ventilation ducts.

2 Crossover point of ventilation ducts with wall penetrations. Distribution circuits of heating pipes with shut-off valves and pumps.

As a rule of thumb, the standard for new buildings requires an installed heating output of 0.1 W/m² when modern high efficiency pumps are used. This compares to about 1.0 W/m² for existing buildings. Through shorter operating periods and demand-controlled pump speeds, it is possible to reduce the demand for secondary electricity by a factor of 15. This means a reduction in electricity demand from 750 to 50 kWh/a for a single family dwelling with 150 m² of floor space.
Replacing conventional pumps with time-switched pumps in existing buildings is an efficient method for significantly reducing the energy demand. By comparison, the insulation of, and in particular, the replacement of old pipe systems, is an expensive measure.

Generally it is recommended that high efficiency pumps of Class A (Energy Efficiency Index EEI <0.4, from 2013 <0.27, from 2015 <0.23) be installed; these should be equipped with a demand-controlled speed controller and timing device.

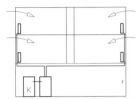

Controlled window ventilation without mechanical ventilation. Intake air is not heated or cooled. Heat input only from radiators.

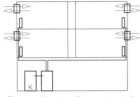

Decentralized ventilation units with heat recovery integrated in the facade. Intake air is preheated by waste heat but is still colder than room air. No pipe network required.

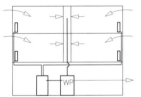

Installation of an air extraction system with heat pump using waste heat. The waste heat is conducted to storage and then utilized in a separate system.

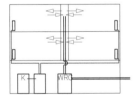

A central ventilation system with air intake and air discharge allows the installation of a heat recovery unit as well as centralized filtering and air conditioning.

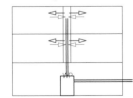

If the energy standard of a building is high, the entire heating requirement can often be covered by heating the intake air. No water-based heat distribution pipes are needed.

Owing to the higher specific storage capacity of water compared to air, water-based heat transport using pumps is significantly more efficient than air based transport using fans. Therefore the volumetric flow of mechanical ventilation systems should be limited to the minimum required fresh air change. The main task of mechanical ventilation is to ensure good air quality rather than to heat/cool the room. Other heat transfer systems should only be omitted when the required air change is also sufficient for the required heating/cooling. From the energy conservation point of view, it is not desirable to recirculate air for the purpose of heating or cooling alone. It follows that the purpose of a room is a decisive factor for the selection of a heat transfer system: densely occupied rooms such as classrooms, theaters, seminar rooms, and event rooms require high volumetric air flows and are therefore suitable for heating/cooling by air (assuming a high standard of building construction); if however only a few persons are in the room, it is preferable to install a water-based heating/cooling system, depending on the amount of heat generated by electrical equipment.

The efficiency of ventilation systems is evaluated in accordance with DIN EN 13779, using the specific fan output (SFP). Pressure loss data can be used to ensure that a system achieves efficiency class SFP1 in operation. The efficiency of the ventilation ducting can be determined by measuring the pressure drop in the duct network. Systems with high pressure losses require greater fan output, which in turn significantly increases the amount of energy used for operating the system.
Pressure drops may be caused by:
- overlong distances
- cross-sections which are too small
- unfavorable shapes (e.g. rectangular instead of round)
- constrictions (e.g. at junctures and feed-throughs)

We can see from the above that ventilation duct networks should be as short as possible, adequately sized, and suitably laid out. Systems which have been optimized from the flow point of view can drastically reduce the amount of electricity consumed by its fans. System designs requiring a high level of maintenance (fire dampers) should be avoided. Therefore it is advisable that the fire protection concept be taken into account at the preliminary design stage and that the duct layout and overflow openings be arranged accordingly. In school refurbishments it is recommended that individual intake and discharge air ducts be installed from each classroom through the floors to the central unit, as this reduces the number of noise attenuators and fire dampers needed. The ducts should be installed all together in an F90 fire resistant enclosure.

Ventilation ducts should be insulated to various levels, depending on the internal and external temperature (thermal conductivity class WLG 040).

Thermal envelope	External air	Exhaust air	Intake air	Exhaust air
inside	100 mm	100 mm	30 mm	30 mm
outside	25 mm	25 mm	80 mm	80 mm

It should go without saying that mechanical ventilation systems are also fitted with heat recovery from the discharge air; in housing, heat recovery may even be the reason why such a system is installed. The EnEV 2009 regulations make it a mandatory requirement that systems with volumetric flows of over 4,000 m³/h be fitted with heat recovery. In addition to the components, the regulations stipulate regular system servicing and inspection. The interval of such servicing depends on how the system is used, i.e. its service hours and the functions it performs (e.g. humidity recovery). For residential ventilation systems it suffices to carry out regular visual controls and, if appropriate, microbiological surface tests. Where acute health problems exist, testing should be performed for airborne germs.

Generally, all components of a ventilation system should consist of material that can be cleaned without damage. Any condensate that may form where the air is treated must be able to drain off unhindered in order to prevent the growth of germs. In addition, filters should be fitted both at the external and internal air intake points, as well as inside the ventilation unit. Filters are grouped into the following classes, ranging from coarse to fine:
- coarse filters, G1 to G4, for dust and pre-filters
- fine filters, F5 to F9, for pollen and main filters
- HEPA filters, H10 to H14, for operating rooms and laboratories
- ULPA filters, U15 to U17, for clean rooms (chip manufacture).

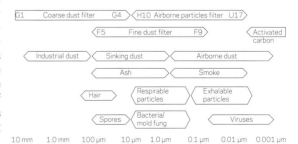

THERMAL AND ELECTRICITY STORAGE

Storage facilities are needed to make supply independent of demand. There are short- and long-term storage facilities, depending on the time period to be covered. Day storage is used for bridging bad weather periods or to cover peak demand at certain times of the day. A long term storage facility stores energy from one season to the next and is therefore also called seasonal storage. Buffer storage is used to optimize the operation of generating devices and is an important component of load management. Important storage parameters are storage capacity, input and output performance, and storage cycles. Heat storage facilities are divided in accordance with principles of physics, i.e. sensible, latent, and chemical (thermochemical) heat storage.

SENSIBLE HEAT STORAGE

When heat is stored in a material, its temperature will always increase. Materials with a high specific thermal capacity are particularly suitable as storage media. Only at each complete cycle of charging and discharging is energy replaced from another source. A fully charged storage facility does not produce any savings if nothing or only very little is taken out of it. A storage facility which covers a day's consumption will be used repeatedly and therefore provides a continual savings benefit. This typically applies to buffer or daily hot water storage with just a few hundred liters volume, for example a tank used in a thermal solar system.

By contrast, seasonal storage only achieves one complete charging cycle per year based on its entire capacity. Such storage only makes commercial sense where its installation means that other services installations can be omitted completely. Long-term storage facilities for sensible heat have to be relatively large since this makes it possible to minimize losses through the exterior surface. Different options exist for achieving such storage. Hot water storage could be in the form of a tank or an underground basin. Underground basins are placed in pits with insulated walls and waterproof covers; these were further developed in the form of shingle water heat storage. Although these require twice the volume, they have the advantage that locally available and cleaned shingle can be used. Where the flow speed of ground water is slow, it is possible to use a ground loop as storage which uses the surrounding earth or rock to store heat. In this case, excess heat is pumped into the ground loop which heats the surrounding soil through heat conduction. Another option is to store heat in an aquifer. In this system, water with excess heat is transferred into the groundwater through one borehole and cooling water is taken from another borehole, which must be between 50 to 300 meters from the first. For this type of system to work, the ground must have certain properties, including those of porosity and water permeability.

LATENT THERMAL HEAT STORAGE

Material can absorb energy through a change in structure. In this case, the storage medium changes phase, typically from solid to liquid and vice versa. Commercially available latent heat stores usually use special salts or paraffins as storage media. These materials do not increase in temperature when they are heated, but melt instead. When new energy is to be stored, it is necessary for the store to thermally discharge first. In this process the medium will recrystallize. When selecting the material for such a system, account needs to be taken of the different volumes involved in changing the aggregate state and the temperature at which this process is supposed to take place. In addition to the latent heat storage some sensible heat storage will also take place. However, it is rather minor compared to the latent storage and therefore of secondary importance. Latent heat storage materials are also called phase change materials (PCMs) and are used for building component activation in lightweight buildings. Typical materials used are e.g. paraffins, which are installed in suspended ceilings and counteract excessive cooling or heating of the interior.

THERMOCHEMICAL STORAGE

In these systems, heat is stored through reversible chemical material reactions and the quantity of thermal energy stored is four to five times as large as that of a conventional sensible hot water store. Suitable reversible reactions in the low temperature range include sorption processes. For example, a sorption store charges as water is extracted from the storage medium. When the process is reversed by exposing the sorption material to water vapor, the bond energy is released as heat. In theory, this process can be reversed any number of times.

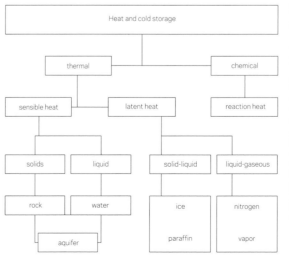

As a comparison and benchmark for the energy density of the different storage systems, the theoretical storage capacity of a hot water tank is calculated here.

The average specific heat capacity of water is 4.2 kJ per liter and degree of temperature difference (corresponds approximately to 1.2 Wh/l*K). The hot water in a tank, which fluctuates between 40 and 90 °C, i.e. with a temperature difference of 50 °C between fully charged and discharged (in relation to the desired temperature level), can thus store 210 kJ of energy per liter (60 Wh/l).

One liter of plant oil (weight: approx. 0.9 kg) contains 33 MJ of energy, i.e. 33,000 kJ. In order to save the equivalent of one liter of heating oil, the tank therefore needs to have a volume of about 160 liters for it to be able to supply the same amount of energy in one complete discharge cycle. It follows that a tank with a volume of 300 liters, as is typical in housing, can theoretically replace nearly 2 liters of oil for each discharge cycle. The critical question is how often this tank will complete a full charging cycle per year.

Solar storage tanks are a form of sensible hot water storage and have to be sized to suit the collector area. Where used for domestic hot water only, a 300 liter tank is common. Installations used to supplement space heating should have a tank with at least 50 liters storage volume per m² of collector surface. A 4-person household with a typical collector area of 12–15 m² should have a storage tank with a 700 to 1,000 liter capacity. As tanks increase in size, their specific heat loss is reduced.

Aquifers or gravel/water systems are used for seasonal storage. These devices only have half the energy density compared to hot water tanks (see table on p.105); therefore the storage capacity has to be increased in volume or mass to suit the respective storage requirement. The typical volume of seasonal storage devices is 20–40 m³ per residential unit. Seasonal stores with a volume of less than 500 m³ are not viable, so it is common to install large facilities providing solar district heating for a whole neighborhood, often with more than 5,000 m³ volume.

Cold stores have to be designed to suit the load profile of the use as well as the cooling plant. Each m³ of ice storage can replace 10-15 kW of cooling output.

Fuels such as wood or oil are really only multi-functional energy stores—the only difference being that they can be "discharged" only once. On the other hand, this discharge does not require high technology - a simple furnace is sufficient. Heat storage refers to equipment such as hot water tanks, which can be charged and discharged indefinitely. When one compares the energy density of such systems with that of fuel, it becomes clear why energy storage is considered the "big technological gap" of renewables-based energy supply. Storage facilities are needed to create a buffer between supply and demand. This will only function satisfactorily if the capacity is adequate. The energy density of a lithium ion accumulator or that of simple plant oil is different by a factor of no less than 100— 1 kg of oil contains more than 100 times the energy that can be stored in 1 kg of an LI accumulator. It is therefore clear that, for seasonal storage, such accumulators do not present a viable alternative to an oil tank, not even in combination with a heat pump.

Energy density in kWh/kg

Lead battery	0.03
Ni-Cd battery	0.04
Ni-MH battery	0.06
Li-ion battery	0.10
Li-polymer battery	0.15
Hydrogen, incl. tank	0.33
Heating water by 1 K	0.0012
Heating water by 50 K	0.06
Melting heat water ..equiv. to approx. 77 K	0.09
Water evaporation ..equiv. to approx. 540 K	0.63
European timber	3.6-5.6
Natural gas (methane)	6.7
Anthracite	8.3
Plant oil	10.3
Heating oil	11.1-12.0
Atomic hydrogen	60
Uranium-235	25 M
Proton-proton fusion	174 M

POTABLE HOT WATER STORAGE

Buffer stores are closed systems that are charged and discharged through heat exchangers. In potable hot water stores the drinking water itself is the heat storage medium. The hygiene requirements are accordingly strict in order to avoid the formation of *Legionella*. The growth of these germs is prevented at temperatures above 60 °C, but this temperature also reduces the storage capacity. The storage volume should be designed to meet the demand and to ensure regular replenishment with fresh water. Potable hot water storage systems include small solar thermal stores as well as under-sink units, whereas instantaneous water heaters are fresh water systems. Small instantaneous water heaters operating at 3-5 kW can be economical for remote showers that are only used occasionally, as these avoid long pipe runs and standing losses. Under-sink units are not recommended due to high losses during periods of no demand.

COLD STORAGE

Like heat, cold can be held in sensible, latent, or chemical storage. At a room temperature of 20 °C, water can be heated in a buffer store by 80 °C (sensible heat storage) without vaporization, but it can only be cooled by 20 °C before it freezes. This means the potential for sensible cold storage based on water is limited and that latent cold storage is more relevant. Ice has a very high storage capacity and can change its aggregate state any number of times. Water is not poisonous and can be used as a storage medium to provide cooling. The melting process takes up the same amount of heat as a temperature increase of 77 °C. So a latent ice store has slightly more storage capacity than a hot water store. In direct systems the cooling agent is also the storage medium, whereas indirect systems involve a heat exchange device.

Cold storage can also be used to balance out fluctuations in demand. They are charged primarily at night. During the day they can cover partial loads completely and reduce the load to be provided by the cooling aggregate at peak periods. This means that the cooling aggregate can be smaller and it also benefits from cheaper electricity tariffs during night-time operation. Where adsorption refrigeration machines are used with an intermittently operating sorption process, their regeneration requires an additional hot water buffer tank, which could be several cubic meters in size. Clearly it is necessary to optimize the system by balancing the heat and cooling generation components as well as the respective thermal storage.

ELECTRICITY STORAGE

A frequent argument used against renewable energies is that the sun and wind are not constantly available and

Type of storage	Energy density [kWh/m³]	Store/ storage medium	Working temperature/ transformation temp.
Sensible	15-30	ground loop store	< 60 °C
	20-30	concrete	< 500 °C
	30-40	aquifer store	< 60 °C
	30-50	gravel/water store	< 100 °C
	60-80	hot water tank	< 100 °C
Latent	> 90	water (solid/liquid)	0 °C
	> 100	paraffins (PCM)	10-60 °C (23-26 °C)
	> 120	salt hydrates	30-80 °C
Thermo-chemical	200-500	metal hydrides	280-500 °C
	200-400	silica gels	40-100 °C
	200-400	zeolites	100-300 °C
	> 10,000	heating oil combustion	> 500 °C

can only be predicted to a certain degree. Whilst it is relatively easy to store heat, it is technically much more complex to store electricity. In Germany, pump-storage hydroelectric power plants assume an important role in electricity supply management. This term refers to power stations that, at a time of excess supply of electricity (low wholesale price), pump water from a lower to a higher basin and thereby store the electricity as potential energy. When demand exceeds supply, this potential energy is converted back into electricity by turbines, just as in an ordinary hydroelectric power station. The reduction in effectiveness is relatively small at around 20%.

Much larger capacity hydroelectric power stations are found in Norway. Further development of the European high-voltage direct current electricity network will enable the utilization of excess production from wind parks during periods of strong winds so that the hydroelectric power stations can be switched off and started up again later when demand exceeds supply, thus acting as electricity stores. These power stations could also be upgraded to full pump-storage hydroelectric power plants by the installation of additional pumps and turbines. With an output of 60 GW, this corresponds to over 40 nuclear power stations. In this way, Norway could become Europe's "battery." Other visionary concepts focus on large solar thermal power stations in the North African deserts, which could also supply Europe with electricity.

GENERATING AND PROCESSING

Boilers use fuel to convert chemically stored energy into thermal energy in the combustion process. Gas and oil boilers are very common and solid fuel boilers for wood pellets are gaining in popularity. Boilers vary according to the fuel they use, but also in their construction and resulting efficiency.

CONSTANT, LOW-TEMPERATURE, AND CONDENSING BOILERS

The term *constant temperature boiler* identifies boilers built up until 1980 that are "control resistant": they are operated at a constant high temperature of 70–110 °C without adjusting to changes in the demand for heat. With an efficiency of frequently less than 70% (relating to the lower calorific value of the fuel, see text in margin) and high waste gas losses of over 10%, this type of boiler is today considered inefficient and not ecological.

Low-temperature boilers (LT boilers) are operated at low flow temperatures of between 45 and 75 °C and can be continually adjusted, depending on the outside temperature. During periods of low heat demand (mild outside temperatures), the flow temperature is kept low and will not rise until the demand for heat increases on colder days. This drastically reduces radiation and cooling losses and such boilers can achieve efficiencies of about 95%. However, the exhaust gas loss is still relatively high at about 8%. With a flow temperature of max. 70 °C and a return temperature of max. 50 °C, LT boilers can also be used in existing buildings, even where the insulation of the building envelope is poor.

Condensing boilers make use of the condensation heat of the water vapor contained in the exhaust gas. In this way they achieve efficiencies in excess of 100%. In these boilers the exhaust gas is conducted past cooler heat exchange surfaces, which are connected to the return pipe; as the water vapor cools down to below its dew point it condenses. Depending on the composition of the fuel (ratio of carbon to hydrogen), its calorific value varies. For oil it is 6%, for gas 12%. While the exhaust gas temperature in LT boilers is still around 120 °C, this can be reduced to 40 °C in condensing boilers. On the downside there is the fact that the exhaust gas discharge is more complicated. Exhaust gas has to be discharged with the aid of fans as it is not hot enough to rise through natural convection. In addition, both flue and boiler may suffer corrosion from the combustion particles retained in the condensate (sooting, pitting). These effects are not critical where gas is used as a fuel, but with oil boilers adequate corrosion protection must be ensured. It is possible to use LT boilers in combination with air exhaust systems and hence condensing technology, while at the same time meeting the higher requirements for flues. The installation uses a pipe-in-pipe system, which is inserted into the chimney. The inner pipe carries the exhaust gas while the surrounding outer pipe is used to conduct fresh air to the boiler. Through the heat exchange that takes place, it is possible to recover not only the exhaust heat, but also the latent heat.

SOLID FUEL BOILERS AND GAS BOILERS

Solid fuel boilers are "omnivores" and can be used for burning almost all solid fuels, especially pellets, woodchips, and split wood logs, but also anthracite and coal. Intensive burning of wood is considered to be a significant source of fine dust, at least compared to gas- and oil-fired, which are low in emissions. Simple room heaters must be viewed critically as their combustion process is not well controlled; by contrast, modern wood pellet boilers meet higher standards. Boilers for wood materials should comply with the (fine) dust emission limits of environmental grade 1 (as specified in the *Blaue Engel* criteria for environmental compatibility); this means a dust concentration of less than 25 mg/m³ in the exhaust gas. It is desirable to achieve half this value, particularly in larger plants.

These boilers are often located in boiler rooms in the cellar with exhaust gases being discharged through a chimney. Domestic gas boilers are compact units that can be installed on walls and used for preparing domestic hot water as well as providing space heating. Their output is usually limited to a few kW and they are therefore suitable for apartments or small buildings. They are usually installed in small rooms, kitchens, or bathrooms, often in combination with a balanced flue system. Fan-assisted exhaust gas discharge of condensing boilers makes it possible to install these in boiler rooms in the attic. Both systems save useful space.

REPLACING BURNERS AND INVESTMENT COSTS

A boiler comprises not only the burner and exhaust gas system, but also a combustion chamber, an insulated casing with a small buffer store, and controls. Replacing a burner is really a repair measure that only makes a limited contribution to improving the efficiency of the boiler. Boiler replacement should always be considered where there is a reason to replace the burner due to poor emission values. Where there is an existing heat distribution system and chimney, heating boilers represent very economical generating devices and only add an insignificant additional cost per kW for larger output systems. Therefore they are also often used in combination with other generating devices to cover peak loads. ⌐ 1 - p.115

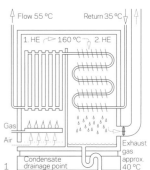

1   Condensing boiler.
Burner (left)
Exhaust gas heat exchanger
(1st and 2nd HE)
Condensate draining point (bottom)
Exhaust gas discharge (right)
Flow and return

2   Exhaust gas heat exchanger.

Traditionally, the (lower) calorific value is used as the 100 percent value which, by definition, does not take into account the condensation heat of the exhaust gases. With the advent of condensing technology, this approach is no longer suitable. Modern units increase the efficiency by lowering the temperature of the exhaust gas and recovering the condensation heat. In this way, they exploit the gross calorific value of a fuel whereas conventional systems only use the net value.

Air collector
Efficiency 60-65%
Operating temperature 25-50 °C

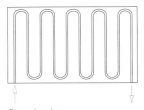

Open absorber
Efficiency 40% (ΔT = 20 °C)
Operating temperature 30-50 °C

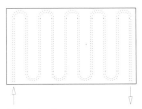

Flat plate collector
Efficiency 65 (ΔT = 50 °C)
Operating temperature 60-90 °C

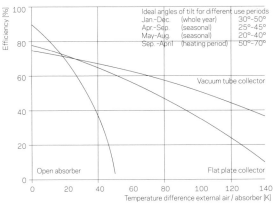

Vacuum tube collector
Efficiency 65% (ΔT = 50 °C)
Operating temperature 70-130 °C

Thermal solar collectors convert the radiation energy of sunlight into heat. This is conducted in a carrier medium to a buffer store or directly to the point of use. There are big differences between various collectors in terms of efficiency, operating temperature, and specific costs. Different constructions are suitable for different locations.

AIR COLLECTORS
Air collectors manage without water as a carrier medium. Fresh or circulation air is conducted through channels behind black-coated metal sheeting and is then conducted directly or through underfloor heating ducts to the rooms. They support the ventilation system of a building. Where they are integrated in south-facing facades in southern countries, they can also replace a heating system: even with moderate solar radiation, temperatures of 25 °C are achieved. Air collectors cannot freeze. On the other hand, air has a lower thermal capacity.

WATER COLLECTORS
The simplest construction form is the open absorber. These absorbers consist of black hoses without any glazing and are often used for heating open-air swimming pools. They are extremely economical and yet can achieve temperatures in excess of 30 °C, particularly in early summer when a moderate increase in temperature is desirable. As the pool water itself is run through the hoses, no additional thermal carrier medium, heat exchanger, or heat store is required. This does not apply in the case of flat collectors. In these collectors, the absorber is surrounded by a box that is covered by glass and backed by insulation. This reduces the amount of heat lost through radiation. As these collectors work with a higher operating temperature, they can be used in a wider range of applications. They are particularly suitable for heating domestic hot water. With a collector area of 1.0-1.5 m² per person, it is possible to cover 50-70% of annual demand. Where it is intended to support the space heating system, three times the area is necessary. The amount of solar radiation available in winter over longer periods is not enough to heat a flat collector sufficiently above the low ambient temperatures. Insufficient heat is generated to warrant its transport; therefore the circuit does not operate in these conditions and in any case it may have to be drained in order to protect it from frost. In this situation, vacuum tube collectors are advantageous. They consist of a row of evacuated glass cylinders containing a selectively coated absorber. The vacuum inside the tube insulates the absorber. Therefore the operating temperature is higher and can also be achieved during adverse external conditions. This means that this type of collector can be used to produce industrial process heat as well as heat for space heating. With collector ar-

eas of about 3.0 m² per person, it is possible to cover over 80% of the domestic hot water requirement and contribute up to 15% towards space heating. However, due to the high initial expenditure on collectors, piping, pumps, solar storage tank, and control equipment such solar thermal installations usually have a payback period of 15 years or longer.

HYBRID COLLECTORS AND PHOTOVOLTAICS
Solar thermal installations are technically simple; nevertheless they achieve high degrees of efficiency. On the other hand, they only generate heat, which is a lower form of energy. Photovoltaic installations generate electrical energy, which is a higher form of energy, but only achieve degrees of efficiency of 5-15%. For this reason it is sensible to combine both these technologies in one collector: while an externally fitted photocell generates electricity, the absorber behind it absorbs the excess heat of the cell as it heats up. Although this principle appears logical, it is technically not easy to implement. The output of a photovoltaic cell decreases with increasing temperature since the electrical resistance increases. For this reason, photovoltaic cells are usually ventilated at the back and they avoid the absorption of those parts of the light radiation spectrum which are not suitable for generating electricity. However, these are of interest for the solar thermal system. So the task is to optimize a combined system which has inherently contradictory strategies. It is therefore not surprising that only a few products have appeared on the market. Roofs should be constructed in such a way that it is possible to subsequently install collector systems. This means that imposed loads of 50 to 100 kg/m² of floor area should be allowed for. Further preparatory provisions can be made in the form of suitable ducts and/or conduits. → 2 - p.115

DISTRICT HEATING SYSTEMS

District heating networks work on the basis that heat is brought into a building from an external source. As with electricity, the building itself does not generate the energy but receives energy in a usable form. Sometimes a distinction is made between larger scale and more localized district heating systems but the principle is the same in both cases. There are several advantages to "outsourcing" the generation of heat: fewer and less technical installations are needed in the building, operational safety is improved, and maintenance is simplified and carried out professionally. For these reasons district heating is often contracted as an external service for large residential developments and public buildings. The degree of sustainability of this type of services concept depends largely on the source of the transported heat, in other words, the primary energy source used for its generation. In many cases remotely produced heat is typically waste heat from combined heat and power stations. In cities, waste incineration power stations are typical, whereas in rural areas biomass power stations are more common. Other sources of waste heat are various kinds of industrial production processes. All these versions can be considered as very environmentally benign as the degree of utilization of the primary energy is increased by utilizing the waste heat generated, provided no additional fuel has to be burnt. Limiting factors are the installation cost and heat loss from the pipe network. Both these factors increase the further the consumer is located from the source.

PIPE NETWORK AND SERVICE CONNECTIONS

Heat is transported through a thermally insulated pipe system, which is usually installed underground and less commonly above ground. Heat carriers are hot water or, in pressure pipelines, superheated water and steam. The specific heat loss of the system decreases with increasing pipe diameter and insulation standards, and lower temperature levels. Typical values are between 5 and 20 W/m. The heat loss varies depending on the quantity of heat transported. If 10% of heat loss from the network is considered acceptable, it is possible to cover a distance of nearly 2 km with 150 mm diameter pipes and a normal insulation standard, but only 1 km with a pipe diameter of 100 mm. In buildings, the service is connected at a transfer station, also called a compact (district heating) station. In a modern installation this is based on a plate heat exchanger. The size of this heat exchanger is rarely determined by the requirement for space heating as the demand for heating domestic hot water (DHW) is more critical. The required service output of such an installation can therefore often be reduced by providing additional domestic hot water storage devices, particularly where the demand for DHW is subject to strong fluctua-

tions. District heat is charged on the basis of the heat delivered; this is measured with heat meters, which record the flow quantity and temperature of the heat carrier medium and hence the transported quantity of energy. The operator of the system calculates the tariff on the basis of his energy costs and the cost of operating the system including any losses. Usually the price of heat is higher, but nevertheless this form of service can be economical, especially for low-energy houses, due to the lower investment cost and the fact that less floor space is needed for installations: these buildings require no heat-generating device, chimney, or fuel storage facility, thus changing the building design quite significantly.

In addition to heating networks, it is also possible to install cooling networks. Networks are grouped into different categories depending on their operating temperature:
- high temperature heat (60 to 130 °C at 16–25 bar)
- low temperature heat (40 to 60 °C)
- very low temperature heat (20 to 40 °C)
- cooling for air conditioning (+3 to +15 °C)
- commercial refrigeration (-37 to +7 °C)
- low temperature refrigeration (< -35 °C)

Depending on the temperature level, a service connection is suitable for generating domestic hot water and for heating or cooling buildings, through to the provision of process heating or cooling. In the case of very low temperature heat, the pipe supplies solar thermal heat, which is further heated with the help of heat pumps.

LOCAL SOLAR DISTRICT HEATING NETWORKS

Germany's Renewable Energies Heat Act stipulates that in new buildings, no more than 15% of the heat requirement be covered by solar thermal generation. Where it is intended to cover significantly more than 30% of the heat requirement with solar energy, a sizeable part of the space heating requirement in winter has to be harnessed from solar radiation. To achieve this, seasonal long-term heat storage facilities are an inevitable prerequisite. They can be used to store part of the heat harnessed from solar radiation during the summer to be used in winter. Larger stores are more compact in terms of their surface area to volume ratio and therefore have correspondingly lower unit heat losses and costs. It is therefore more economical to supply heat to a whole neighborhood with a local district heating network from one large central storage facility. If 0.25 m² of collector area and 0.4 m³ of storage volume are provided per m² of net dwelling area, it is possible to cover more than 50% of the heat requirement. In order to guarantee a reliable supply in bad weather periods, it is necessary to provide a backup boiler. ⤳ 3 - p.115

1   District heating pipe segments.

2   Local solar district heating neighborhood.

3   Seasonal storage facility.

District heating installations require storage facilities and pipe networks outside buildings, but on the other hand, they save space for technical equipment inside buildings. Transfer stations and buffer stores only take up a few square meters of space and do not require special provisions for fire protection and ventilation.

A large modern coal power station achieves a maximum efficiency of 45%. The average efficiency of large power stations in Germany is about 32%. That means 55–68% of the energy is released in the form of waste heat, and is therefore lost. With large power stations, the consumers of heat are usually far from the plant (where the heat is generated), so using the heat for district heating networks would thus involve significant heat losses in transport, expensive pipe networks, and high input of secondary energy for operating pumps.

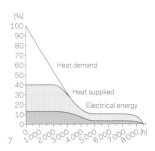

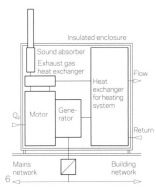

4   Combined heat and power (CHP) station at Frankfurt am Main.

5   Modular CHP on anti-vibration pads and with sound insulation enclosure, SMA production building.

6   Diagram of a modular CHP unit.

7   Heat and electricity supplied by a modular CHP unit.

Combined heat and power (CHP) refers to the simultaneous generation of heat and power. Instead of burning fuel directly and exclusively for generating heat, a CHP system generates electricity with the help of a motor-driven generator. The waste heat generated in this process in the exhaust gas and cooling water can be utilized for covering the heat demand of those consumers connected to the system. In combined heat and power generation, the energy content of the fuel used is not utilized any better than in modern condensing boilers. However, CHP produces electricity, which is a higher form of energy and can be used for a variety of purposes. Small CHP systems can also be used locally to generate heat and power directly at the consumer's premises, so that it is not necessary to install a heat distribution network with its associated distribution losses. Such decentralized CHP systems achieve a significantly higher degree of efficiency (net energy electricity plus heat divided by primary energy used) compared to the conventional system of using localized heating for the heat requirement and central electricity distribution (with losses of up to 70%). They are therefore a suitable means for increasing efficiency.

MODULAR CHP UNITS

Increasingly popular versions of CHP stations are sometimes called *modular CHP* units. These are motor-based CHP units for small- to medium-sized applications. They include Micro-CHP units, which are available with electric outputs from 1 to 5 kW and waste heat outputs of from 4 to 12 kW. Apart from the output, the most important rating factor for modular CHP units is the ratio between power and heat generation. It is referred to as the electrical output modulating range and reduces with smaller systems (1 kW$_{elec}$: 4 kW$_{therm}$ = 0.25, 5 : 12 = 0.42). In addition to internal combustion engines, other systems for power generation are in use or undergoing trials, such as Stirling engines, fuel cells, micro turbines, or steam engines. Possible fuels are primarily fossil fuels and regenerative hydrocarbons such as heating oil, plant oil, and biodiesel (for diesel engines) or natural gas/biogas (for gasoline engines and gas turbines). Fuels such as woodchips and wood pellets require external combustion (Stirling engines and steam power plants) or have to be converted to gas beforehand. These two systems are not commercially viable unless for larger plants.

HEAT, ELECTRICITY AND NETWORK-BASED MODULAR CHP UNITS

There are three different types of modular CHP. Electricity-based systems are designed to meet local electricity demand, whereas heat and network-based units are designed to meet the heat demand. In systems laid out to cover a certain electricity demand, any excess heat generated that cannot be used at the time has to be kept in a buffer store for later use, or has to be fed in to a local district heating network. Only in exceptional cases should excess heat be given off to the environment through an emergency cooling device. In the latter case, the overall degree of efficiency is no higher than that of a large conventional power station that does not harness the heat output. In systems that are designed to cover a certain heat demand, any excess electricity is fed in to the mains network. In that case, the overall degree of efficiency is always high. Heat-based modular CHP systems are usually designed such that their output only covers part of the maximum heating energy demand of the connected consumers, even when operating at full load. In this way the modular CHP is designed as a base load generator, and peak loads in winter are covered through other, more economical generators, such as boilers. Separating base load and peak load generation is a way of ensuring that the expensive electricity generating systems are utilized to maximum efficiency. The design of these units should aim at a minimum of 3,000 (preferably 5,000) operating hours out of a total of 8,760 h/a available. Network-based systems are controlled remotely by the utility company. This company will use the many existing decentralized small power stations to manage the load on the electricity network. Excess heat is fed in to a local buffer store from which the building can draw heat as and when required. Where an emergency power supply is required (for example, in hospitals to ensure supply in the case of a power cut), it is preferable to provide this from a modular CHP, unless this is undesirable for operational reasons.

COMBINED COOLING, HEATING, AND POWER

Combined cooling, heating, and power (CCHP) is an extension of the principle of combined heat and power: the heat generated by a modular CHP, which is used in winter to heat buildings and in summer to operate the absorption chiller of an air conditioning system. In spite of the fact that the absorption chiller is considerably more expensive (compared to an electrically powered compression chiller) and is more difficult to integrate technically, CCHP has certain advantages. In summer the heat generated by the unit is available for the preparation of domestic hot water and, in addition, provides the heat required by the evaporator of the absorption chiller. This makes it possible to increase the size of the modular CHP, significantly increase the number of hours it is in full operation during a year, increase the amount of electricity generated locally, and improve payback. ⌐ 4 - p.115

HEAT PUMPS

Heat pumps are used to exploit environmental heat at relatively low temperature as an energy source for buildings with the help of a heat carrier medium. The circulation process is based on the same principle as a refrigerator, the difference being that in a heat pump, the medium coming from the outside of the building is cooled down in order to provide heat inside the building. The heat carrier in the heat pump changes its aggregate state (liquid or gaseous) depending on the pressure and temperature conditions in the circuit, and thereby absorbs or gives off heat. The heat provided by heat pumps is at a temperature level that is primarily suitable for heating systems with low flow temperatures. In existing buildings and in the preparation of domestic hot water, the efficiency of heat pumps is reduced. The following heat sources and heat exchange mechanisms are possible:
- ground source heat (retrieved using ground loops, vertical bore holes, foundation piers, earth tube collectors, approx. 75% of market share);
- groundwater (using a well and infiltration sink);
- aquifer (from seasonal storage, retrieval as in groundwater);
- external air, waste air (retrieval using heat exchanger, possibly preheated air from tube collector, air collector, interiors);
- exhaust heat (retrieval using heat exchanger, for example from industrial processes or wastewater).

It is also possible to combine systems in which the heat carrier medium is preheated by air or solar thermal collectors. By increasing the temperature level it is possible to more efficiently convert the environmental heat from the heat pump into usable heat.

Heat pumps can also be distinguished by the type of heat carrier. The first term of the descriptive word pair refers to the circuit outside the building for retrieving environmental heat and the second term refers to the circuit within the building (water- or air-based distribution system):
- water to water and water to air;
- ground to water and ground to air;
- air to water and air to air.

From the point of view of using primary energy, efficient heat pumps are one of the best heating technologies if the operating energy is derived from renewable sources. Where a building has a photovoltaic installation, the electricity generated can be used for operating the heat pump; however it is important that demand and yield are balanced as much as possible for daily and annual consumption. This is made easier with the help of buffer stores on which demand can be made for a few days when needed. If a heat pump is operated by mains electricity it cannot be considered favorable from the primary energy point of view unless the seasonal energy efficiency ratio is significantly higher than the primary energy factor for the operating energy (EnEV 2009: $f_{P, electricity}$ = 2.6).

COEFFICIENT OF PERFORMANCE

Heat pumps in domestic buildings are almost exclusively compression systems. Their compressors are operated with electrical energy; larger systems can also be operated with gas-powered compressors. The *coefficient of performance* (COP) of a heat pump is defined as the ratio of energy input to heat output. When this is based on electrical energy input of a compressor, the COP is calculated as the ratio between the heat transferred to the flow and the electrical energy required by the compressor. A COP of 4.0 means that the available heat is equivalent to four times the electrical energy used. This means that the heat for heating consists of three parts environmental heat and one part electricity. This COP is typical for a ground-to-water heat pump where environmental heat is available at 10 °C and usable heat is to be provided at 50 °C (the code for this type of heat pump performance is B10/W50 COP 4.0). The COP value will decrease with decreasing temperature of the heat source and increasing temperature of the required heat distribution.

SEASONAL ENERGY EFFICIENCY RATIO AND OPERATING MODES

The *seasonal energy efficiency ratio* (SEER) is the coefficient of performance that is actually achieved over the course of an operating year. It is determined by measuring the electricity (meter) to establish the electrical input energy, including auxiliary energy for the compressor and heat source pump, and measuring the thermal output with a heat meter. The SEER is the ratio of the annual heat yield to the expended energy input. The operating mode of a heat pump also significantly affects its cost efficiency. Common operating modes:
- monovalent
(heat pump operation throughout the year);
- mono-energetic
(with an electrical heating rod for cold days);
- bivalent
(supplemented with a peak-load heat generator).

In bivalent operation an additional heat generator is activated when the external temperature drops below a specified value. This temperature is referred to as bivalence temperature. → 5 - p.115

Coefficient of performance: Degree of efficiency of heating or cooling output relative to the mechanical work employed. The degree of efficiency pertains to single points of time with clearly defined conditions.

Heating: coefficient of performance
$COP = Q_H / (Q_H - Q_{ambient})$

Cooling: energy efficiency ratio
$EER = Q_K / (Q_{waste heat} - Q_K)$

Efficiency rating:
A heat pump's efficiency rating (η_{HP}) is the coefficient of performance (COP) actually achieved for the temperature level in use, relative to the ideal COP. Typical heat pump efficiency ratings are between 0.45 and 0.55.

In manufacturers' data sheets, output data are related to the type of heat carrier and specified source and target temperatures.
Examples:
W10/W50
COP 4.5 / η_{HP} 0.56 / 12.0 kW
B10/W35
COP 6.1 / η_{HP} 0.58 / 13.8 kW
B0/W35
COP 4.8 / η_{HP} 0.55 / 10.4 kW
B0/W50
COP 3.6 / η_{HP} 0.56 / 9.0 kW
A10/W35
COP 4.3 / η_{HP} 0.35 / 8.8 kW
A2/W50
COP 2.7 / η_{HP} 0.40 / 6.8 kW
A2/A35
COP 2.9 / η_{HP} 0.42 / 3.5 kW
(where W = water; A = air; B = brine)

Minimum COP values to qualify for the 2011 EHPA quality seal:
water to water heat pump:
W10/W35 - COP 5.1
brine to water heat pump:
B0/W35 - COP 4.3
air to water heat pump:
A2/W35 - COP 3.1

Seasonal energy efficiency ratio (SEER): the coefficient of performance actually achieved over the course of an operating year.
$SEER = Q_H / Q_{drive}$ [annual]

The degree of utilization pertains to a certain period, under varying conditions, and must determined individually for each building, taking into account internal and external factors and the operating mode. SEER is the critical value determining cost efficiency. The SEER must be greater than 2.5 for the cost per kWh of net heat to likely be less than with an ordinary boiler.

1 Adsorption chiller,
SMA production building

Compression: condensing and expanding a refrigerant in a closed loop system.

Adsorption: the refrigerant (water) attaches to a sorption agent, such as silica gel or zeolite.

Absorption: the refrigerant (ammonia or lithium bromide) is blended with the liquid sorption agent (water).

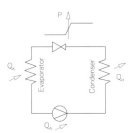

Schematic for compression. Components: compressor (compresses the refrigerant), condenser (releases heat energy), expansion valve (expanding the refrigerant), cooler (absorbs evaporation energy)

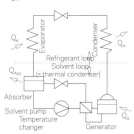

Schematic for adsorption. Components: Absorber (absorbs and liquefies the refrigerant vapor), desorber (separates the refrigerant from the sorption agent), condenser (releases heat energy), cooler (absorbs evaporation energy)

Where less than 40 W/m² of cooling output is required, passive cooling of buildings is usually sufficient. For higher cooling outputs it is necessary to involve active systems in order to guarantee comfort levels in summer. Mechanical chillers work either by compression using electricity only or mainly thermally through sorption. Both systems use identical circulation principles: a refrigerant is evaporated and thereby absorbs ambient heat; this heat is released again when the refrigerant vapor returns to a liquid aggregate state and has to be released to the environment via coolers. The two systems have different operating modes: in sorption systems the refrigerant is "thermally condensed" while in a compression system it is compressed with the help of an electrical compressor.

COMPRESSION CHILLERS

Compression chillers work on the same principle as heat pumps. Highly efficient systems achieve performance coefficients (usable cooling output to input energy) of between 5 and 6; modern standard systems achieve COPs of 3 to 4 and older systems often under 2. As the technical principle is the same as that in a heat pump, these machines can also be used as a reversible heat pump for heating and cooling. A disadvantage is the high degree of maintenance required due to the many moving parts; another disadvantage is the fact that the refrigerant can be detrimental to health and the environment. It is preferable to use only non-halogenated or partially halogenated refrigerants, such as water (R 718), carbon dioxide (R 744), and ammonia (R 717).

SORPTION CHILLERS

Where high temperature waste heat (> 80 °C) is available from combined heat and power stations, industrial processes, vacuum tube collectors, or modular CHP units, sorption chillers are an option. While compression chillers are an option only use electricity, sorption chillers run mostly on heat, which is used to regenerate the used carrier medium and prepare it for recirculation. The systems may use either absorption or adsorption. Absorbents are liquids while adsorbents are solids. Liquids allow continuous operation whereas solids require periodic operation. The latter can be compensated with two units used in phase shift mode or by using a cold water store. Both systems have the same advantages: low environmental impact of the refrigerant and low maintenance and operating costs. The disadvantages are higher investment costs and a greater space requirement. Also the output from the cooling units (heat discharge above roof level) is significantly higher as humidity is evaporated. The thermal performance coefficient ranges from 0.3 to 1.2, and the electrical coefficient from 6 to 10.

EVAPORATION COOLER

The externally visible components of a chiller are the air intakes and evaporation coolers, both of which may be installed on roofs or suspended on facades, and rarely look attractive. They discharge the inevitable waste heat from chillers to the environment at increased temperature levels. Part of the waste heat can also be used for heating domestic hot water although this is only possible where the load profiles are suitable. In open wet cooling systems, the cooling water is mixed with external air in aerating devices within the unit. As evaporation heat dissipates from the water, the remaining water is cooled by 6 °C for each 1% of evaporation. In addition, the water is cooled through convection when it is aerated. In closed wet cooling systems, the water to be cooled is not in direct contact with the external air but is conducted separately through a heat exchanger. This system is slightly more expensive but is preferable from the hygiene point of view. Wet or evaporation chillers are compact and economical; due to their high thermal efficiency they use little energy (electricity used for operating the pumps) but 5% of the water to be cooled will evaporate. In colder external temperatures, it may be possible to omit the evaporation function. In this case the heat discharged through convection is sufficient and the wet cooler is run as a dry cooler.

In dry cooling, heat is discharged to the external air through heat exchangers. In this type of system, it is often necessary to use fans to increase convection. The heat exchanger tubes are fitted with cooling fins in order to increase the transfer surface. The investment and the energy used per unit of output are higher but the advantage is that no water is used. Decentralized split air conditioning units are dry cooler units. In these systems the condenser is placed outside the room to be cooled and is connected to the evaporator inside the room with short hoses. Some split units can be reversed to function as heat pumps, which makes it possible to use the air conditioning unit as an external air source heat pump as an additional source of heating during the colder seasons. Monoblock units are less efficient as they combine both aggregates inside the room and discharge waste heat to the outside through discharge hoses. 6 - p.115

PASSIVE COOLING

Passive cooling systems are designed to discharge moderate heat loads of 20–40 W/m² using natural means. They utilize favorable temperature levels in the air, ground, or water. Passive systems also include those systems where heat is discharged through evaporation. In ideal conditions these systems can be used to achieve comfort levels without mechanical means. But sometimes the natural cooling potential may be limited in output or only available at certain times, or may not be fully controllable, or may depend on certain external factors. In these cases, passive systems may be used to increase comfort levels or pre-cool the air with cost-saving devices. Where comfort levels are confined to strictly specified narrow ranges, it is usually necessary to have fully controllable supplementary mechanical cooling.

GROUND

Owing to its large thermal mass, the ground represents a substantial potential for conditioning the air supply to buildings. Without human interference, the temperature level is constant at increasing depth. At a depth of 30 m, this temperature equates to the average annual temperature, which in Germany is 10–12 °C. Relevant influencing factors are the location, the groundwater level, and the characteristics of the soil. In winter, heat pumps can be used to exploit the heat stored in the ground for heating. Conversely, in summer the soil temperature is low enough to be used for cooling. With a suitably designed system, the alternating withdrawal and discharge of heat in the respective season leads to a temperature differential of 4–6 °C in the ground. In an ideal situation the annual input and output are balanced so that, over the long term, the temperature of the ground does not gradually creep up or down.

The heat transfer system can be based on air or water. Air to air ground heat exchangers are used for pre-cooling/pre-heating mechanically drawn-in external air. They consist of nearly horizontal ducts buried in the ground (pipe diameter from 150 mm) or ground loops (narrow arrays of pipes). As it is possible for condensation to form in the pipe during summer, a slight slope of min. 2% is important in order to ensure that any condensate can drain away. Stagnant water allows microbes and mold to grow, which would lead to contamination with spores and unpleasant odors in the intake air. Usually the lowest point of the system is at the transfer station in the house, where a drain-off facility should be fitted. With long pipe systems, inspection chambers are provided outside the building, and the system is installed with these chambers at the low points. Any condensate is removed by pump; in addition these inspection chambers are used for maintaining and cleaning the network. The pipe material typically consists of plastic piping with a smooth surface; in large systems (with walk-in pipes), concrete and fiber cement pipes are more economical. Such ground loops can be laid in the building pit around the building perimeter, under green areas, and also underneath the ground slab of the building prior to its construction. Some projects also include a complete lower floor designed like a labyrinth with many partition walls. For areas not in contact with groundwater, rain, or sunshine, it is important that the annual input and output are balanced. Where intake air is pre-heated, there is no need for installing a pre-heating device in the ventilation system. Depending on the pipe network size (25–45 m), it is possible to achieve temperature increases of up to 20 °C and cooling of up to 12 °C.

Water to air systems use water as the heat carrier medium; the closed ground loop is connected to the air intake system through a heat exchanger. This allows smaller pipe cross sections and any hygiene problems are avoided. In water to air systems, ground loops can be installed in the same locations as in air to air systems; for air flows of 200 m³/h, the ground pipe needs to be about 100 m long. It is also possible to utilize pile foundations for geothermal purposes, where these piles are required for structural purposes.

GROUNDWATER

The temperature of groundwater varies between 8 and 12 °C, which means that it is ideally suited to heating and cooling. By contrast with ground loops, no performance-reducing heat transfer is required, as water is taken out and returned directly. For this purpose, a well is drilled at a distance of about 15 m from an infiltration sink, both of which reach to 5–6 m below groundwater level.

ADIABATIC COOLING / EVAPORATION COOLING

When water changes from its liquid state into a gaseous state, it absorbs heat. This is taken from the surroundings. In direct adiabatic cooling, the moisture content of the room air is increased. This can be achieved technically by spraying mists of fresh water into the intake air duct of a ventilation system, or naturally from open water surfaces, water walls, fountains, and planted atriums. The principle is very simple, which is an advantage; on the other hand, the direct evaporation causes an increase in relative humidity. This is perceived as an increase in room temperature and is therefore counterproductive. In addition, the effect fails in summer when the external air is already very humid and can no longer absorb any more moisture. For this reason, direct adiabatic cooling is used in situations where an increase in relative humidity causes no problem, or is even advantageous, for example in dry hot climate zones or in buildings with low moisture input.

1 Planted wall.
Caixa Forum, Madrid, 2008,
Herzog & de Meuron.

2 Area of water.
Federal Chancellery, Berlin, 2001,
Axel Schultes Architects.

3 Planted atrium.
Institute building, Wageningen (NL),
1998, Behnisch & Partner.

3

Indirect adiabatic cooling avoids increasing the absolute humidity of the room air, as in this case the moisture content of the exhaust air is increased. The cooling effect in the exhaust duct is then transferred to the intake air through a heat exchanger. The technical equipment required for this type of system includes mechanical air intake and air extraction with heat recovery. The additional cost of indirect adiabatic cooling is solely in the mist spray process. In practice, this system is used in rooms with moderate thermal loads (< 40 W/m²) and in buildings that are not used at night and have mechanical ventilation anyway, such as offices and schools .

AIR HANDLING SYSTEMS
Where the passive measures described above are not sufficient to pre-heat or pre-cool the intake air, a simple ventilation system can be supplemented with additional functions. The overall term is *Heating, Ventilation, and Air Conditioning* (HVAC), which covers functions such as ventilation, heating, cooling, humidification, and dehumidification. Full air conditioning systems are HVAC installations with all the above functions. Compared to systems that use building components for heating and cooling, in particular ceilings and ceiling panels, the degree to which

the intake air can be heated or cooled is limited. Where it is intended not to exceed the number of air changes required for hygiene purposes, 25–50 W/m² can be extracted or supplied, depending on the type of air outlet. Only recirculation makes it possible to achieve a higher output, albeit at the expense of efficiency and quite often also, comfort (drafts, noise). On the other hand, HVAC systems are laid out to control the relative humidity of interiors. In winter, such a system will provide heating and humidification, and in summer cooling and dehumidification.

HEAT AND HUMIDITY RECOVERY

Ventilation systems use electricity to drive the fans but can also achieve a significant reduction in heat demand through heat recovery ventilation (HRV). The efficiency of heat recovery is usually measured in terms of the heat recovery coefficient and the overall heat transfer coefficient. The heat recovery coefficient (also called heat supply rate or temperature change rate) refers to the temperature difference between intake and exhaust air, and exhaust air and external air. The overall heat transfer coefficient not only refers to these temperature differences but also to the heat gain from fans, compressors, and intake air controls, as well as the latent heat content in humid air. In general there are five different construction types, which differ with regards to their heat recovery coefficients:
- cross flow plate heat exchanger (50–70%)
- cross-counter flow plate heat exchanger (70–90%)
- combined circulation heat exchanger (40–70%)
- heat pipe heat exchanger with fins (40–70%)
- rotary heat exchanger (50–80%)

With rotary heat exchangers it is possible to recover heat as well as humidity. A rotating disc transfers heat between the outbound and inbound air. Humidity is transferred through condensation on the disc and the demand for humidification and dehumidification of the intake air is reduced. This system is capable of recovering up to 80% of the heat (heat recovery coefficient and overall heat transfer coefficient). It is imperative to avoid any transfer of odors or harmful substances. In certain applications it is not possible to recover humidity from the exhaust air unless filters are used to satisfy hygiene requirements. For high thermal loads in summer, a temperature gauge controlled bypass should be installed for heat recovery. If the room air is cooler than the external air, heat recovery can make an important contribution to pre-heating, even in summer. If not, the bypass will be used.

AIR CONDITIONING SYSTEMS

Conventional air conditioning systems achieve cooling and dehumidification by passing air over cooled surfaces so that its temperature drops below the dew point. Fresh air is cooled much below the temperature required for intake air so that its relative humidity reaches 100%, at which point condensate precipitates and can be drained off. This means that the chiller has to operate at very low temperatures and hence uses an increased amount of energy. In addition, the intake air has to be reheated prior to distribution in order to reduce the relative humidity from 100% to an agreeable level. The additional energy required by modern air conditioning systems during a cooling cycle depends largely on the efficiency of the cooling circuit. This has steadily improved over the years. High quality air conditioning units achieve coefficients of performance of 3.5 to 4.0 for cooling. Air conditioning systems are also increasingly used for heating—this is because about two thirds of the heat they supply is taken from the external air and only one third is heated electrically. In design they are very similar to (external) air-to-water heat pumps. Modern units can achieve seasonal energy efficiency ratios (SEERs) of up to 5.0 in heating mode.

DEC SYSTEMS

As an alternative to air conditioning based on compression, dehumidification can also be achieved using a sorption process: in this type of system, humidity is adsorbed on to a material rather than drying the air by decreasing its temperature below the dew point. A version of sorption cooling that can be integrated directly in HVAC systems is the DEC system (*desiccative evaporative cooling*). This uses an open adsorption process involving sorptive dehumidification and adiabatic cooling: filtered external air is dehumidified in a sorption regenerator, which releases condensation heat. The dry but warmed air transfers its heat to the outbound air through a heat exchanger. The intake air is further cooled to the required level in a controlled evaporation cooler. To ensure that the exhaust air is initially available as a heat sink in the heat exchanger and immediately thereafter as a heat source in the sorption regenerator, it is first humidified and then, after the heat exchange, reheated again. The investment cost of DEC systems for moderate volumetric flows are lower than those for conventional compression cooling systems. In addition, excess heat is discharged directly into the outbound air flow after the cooling process, so that no cooling units are required. However, the performance of a DEC system soon reaches its limit where the relative humidity and external temperature are high.

Example of heat recovery:
A single family dwelling with an air volume of 500 m³ and a mechanical air change rate of 0.60 per hour requires 300 m³/h of volumetric flow. If there is a 20 °C temperature difference between inside and outside, the heat output from a heat recovery system at a heat recovery rate of 75% is no less than:
300 m³/h ** 0.34 Wh/m³K * 0.75 = 1,530 W = 1.5 kW

This means that over a period of twenty-four hours, nearly 40 kWh of heat is saved; depending on the efficiency of the heating system, this corresponds to 4 or 5 liters of heating oil (or m³ natural gas). The input energy required to achieve this is 1.5 kWh of electricity. In terms of primary energy (electricity 2.6, heating oil, and natural gas 1.1), this represents a cost to benefit ratio of 1:10, i.e. an improvement in comfort and a worthwhile financial benefit.

Passive heat recovery is more beneficial from the energy point of view than the use of a heat pump (active device). Therefore a heat pump should only be used as a secondary measure or in the case of systems based on exhaust air only.

Recommendations for controlling humidity:
The dehumidification of intake air represents a sizeable proportion of the energy requirement in air conditioning systems. Specifications for very narrow ranges of relative humidity can only be satisfied with active systems. This means that even in museums and galleries where there is a need to preserve exhibits, the target humidity and temperature levels should be agreed as a sliding parameter for the different seasons in order to avoid excessive humidification loads. Frequently it is much more important to limit the speed of temperature and humidity changes for the particular use requirements, e.g. to <1% or < 1 °C per day.

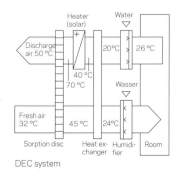

DEC system

The flow diagrams illustrate the chain of losses for different fuels, from the primary energy input (black = fossil, gray = renewable) during generation, storage, and distribution through to consumption as heat for space heating or domestic hot water preparation. All these generation diagrams are based on the same net output energy, but the primary energy input varies greatly, depending on the system. This means the diagrams are generated from right to the left, i.e. from the net energy to the primary energy. Systems producing energy that is fed into the mains network obtain credits.

1 Condensing boiler.

2 Solar thermal installation (with heating support).

3 District heating power station.

4 Modular CHP unit.

5 Heat pump (seasonal energy efficiency ratio: 3.5).

6 Sorption chiller.

Primary energy -> Final energy -> Net energy

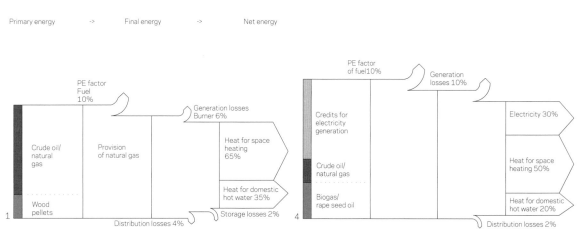

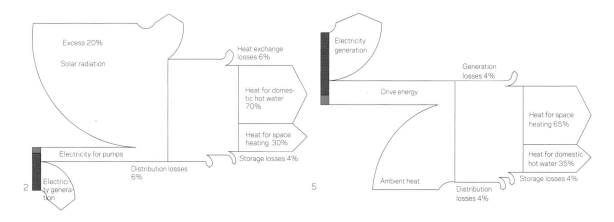

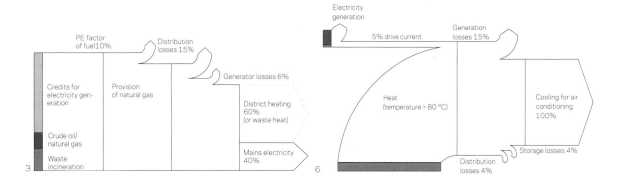

QUALITATIVE SYSTEM DESIGN

Both the use and construction standard of a building contribute to the amount of energy needed over the course of a year. These factors can be used to divide the services to be provided into suitable or necessary system components. In such an analysis, the energy demand can be divided into base and peak load and into the required time periods, and the required flow temperature of the thermal transfer system can be determined, as well as the suitability of locally available energy sources and utilities. Typical system component designs of modern services installations are shown on the following three pages in relation to the annual cycle of energy demand. The issues are dealt with in sequence (from top to bottom):

1. Annual cycle of heat and electricity demand
The monthly energy requirement for domestic hot water (Qww), heating (Qh), cooling (Qc), and electricity (Qelec) is established. The demand is illustrated as unfilled columns, heat demand above the zero line and electricity below. These data have to be researched for each specific use; the Internal Factors chapter provides approximate values.

2. Providing base load energy
Base load energy should be provided by systems using renewable energy with optimum efficiency. As these systems are usually fairly cost intensive in terms of capital cost per kW output, they should be designed to achieve a high degree of utilization in terms of operating hours in order to ensure a reasonable pay-back time. The energy provided by the base load facility is shown as partial filling in the columns for demand, based on its limited nominal output. Diagram 2 also includes solar systems in which the generation of energy cannot be controlled and which is subject to seasonal fluctuations.

3. Providing peak load energy
The remaining energy demand to be satisfied is shown as the complete filling of the columns for demand with a hatched area. Where electricity is used, this is covered by the mains electricity.

4. Proportion of renewable energy sources
Based on the net demand for heating, cooling, and electricity it is possible to derive a final energy demand typical for the system. Taking into account the energy provided by utilities, the proportion of renewable energy is calculated. The green color indicates the proportion of renewable energy sources, and the black color that of fossil fuels. The composition of energy sources can also be used to determine other elements such as the cost of energy and amount of CO_2 emissions.

The total energy demand varies throughout the year, due to seasonal changes as well as holiday periods. The model works on the basis of calculating the energy to be provided in each month. The annual cycle of energy demand provides relevant information to provide an appropriate amount of energy when it is needed, particularly so for renewable energy sources. For the design of the base and heat load generator it can be helpful to order the demand based on amount (see margin). Base load generation systems should be designed to achieve maximum operating hours in the optimum performance range. It is therefore important to avoid oversizing and thereby obviate standstill periods, frequent interruptions of the system or continuous operation at partial load.

The quantity of heat produced by a heat generating appliance is the product of output and time. The quantity of heat to be produced can be graphically determined by entering the continually changing heat load throughout the year on the y axis, and the selected time period on the x axis; the area described by the load curve represents this heat quantity. Any generating appliance can only cover the proportion of energy that can be calculated from its nominal output in kW multiplied by operating hours in h. The arithmetic mean shown in the diagrams (average of all monthly values) provides some initial information for the design. However, it is possible that in the case of very different monthly values (for example, three winter months with a very high demand for heat) the average can result in an inefficiently sized system. For this reason, another method is introduced here instead of the calculated average. Quantiles are suitable means for arriving at an initial qualitative dimensioning of a system.

Quantiles are points at regular intervals along a distribution curve in statistics. Particularly important is the median (also called the 2-quantile which is at the 0.50 interval) and the quartiles (the 4-quantiles at the 0.25, 0.50, and 0.75 intervals). The median or central value divides the entire distribution in two equal parts, while the quartiles divide the distribution into quarters. This means that the median describes a value below the average value, which nevertheless covers 50% of an area. It follows that the median is a possible quantity input for designing a base load generating device.

DIFFERENT TYPES OF GENERATING DEVICES
The different types of generating devices can be classified by the system they employ and by their capacity:
- monofunctional generating devices (boilers, collectors, compression chillers);
- bifunctional alternative generating devices (reversible heat pumps, heat exchangers);
- bifunctional parallel generating devices (combined heat and power, hybrid collectors);
- transformers (sorption chillers, DEC systems);
- combined generating devices (combined heat and power).

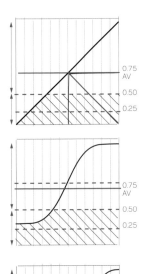

Monthly load profile diagrams.

These three distributions illustrate how the median (2-quantile or 0.50) and also the 4-quantile (or 0.25) can be used for sizing a base load heat generating device, while the average value (MW) would lead to over-sizing.

This is caused by the method with which these two values are calculated: in the calculation of the average value, a few, very high values can lead to a very high result. If the heat generating device is designed for this value, it may not be fully utilized for extensive periods of the year, and yet would not nearly cover the peak loads (see lower diagram, MW curve). By contrast, the median line (0.50) only takes limited account of the individual extreme values.

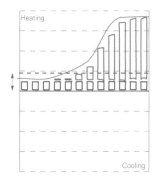

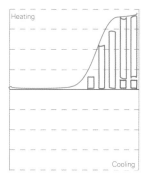

Heating and cooling load diagrams.

RESIDENTIAL BUILDINGS
Heating: solar thermal + condensing boiler
Cooling: none
Electricity: mains current

1. Residential buildings should be designed such that they do not require cooling in a moderate climate. The electricity demand is relatively low and mainly determined by the way the household is run. For this reason the heat required for space heating determines the design of the services installation. With higher building construction standards the proportion of energy required for the preparation of domestic hot water increases; in buildings designed to Passivhaus standard, both types of heat demand are approximately equal. The demand for domestic hot water is largely constant over the course of a year, while the demand for space heating is subject to strong seasonal fluctuations. For new buildings built to EnEV 2009 the peak load in winter is 4-5 times higher than the base load in summer; in passive houses it is only 2-3 times the base load.

2. For new buildings, the Renewable Energies Heat Act (REHA) stipulates the integration of systems based on renewable energy. Installing a solar thermal system for domestic hot water and space heating supplemented with a condensing boiler meets the REHA requirements. It must be said though that solar thermal systems can only cover a limited proportion of total demand. Depending on the size of the collector and its location, the system will produce unused excess during summer while it will not produce any heat during winter. For this reason it is not economical to add additional collector output except where seasonal heat storage is included in the concept.

3. The installation of a solar thermal system means that the back-up condensing boiler can be switched off completely during the summer period, while on the other hand it has to be designed to cope with the maximum heat load in winter. This means that this condensing boiler only operates in certain seasons and achieves full load operation only for a few hundred hours per year, even though it may provide more than 80% of the heat requirement. As the price for the average output of small gas boilers is relatively low, they are suitable for this purpose. Any demand for electricity has to be fully covered from the mains network since no electricity is generated on site.

4. The limited proportion of cover supplied by a solar thermal installation is also clear from the composition of the primary energy requirement. The largest part of the energy requirement is covered by fossil fuels. Even the use of biogas would not significantly change this situation, as the simple combustion of biogas is not desired by legislation.

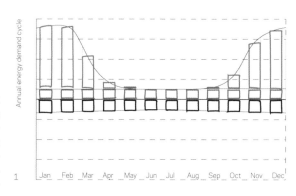

1   Jan   Feb   Mar   Apr   May   Jun   Jul   Aug   Sep   Oct   Nov   Dec

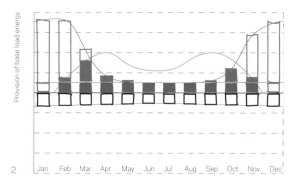

2   Jan   Feb   Mar   Apr   May   Jun   Jul   Aug   Sep   Oct   Nov   Dec

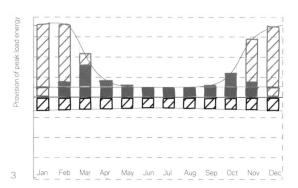

3   Jan   Feb   Mar   Apr   May   Jun   Jul   Aug   Sep   Oct   Nov   Dec

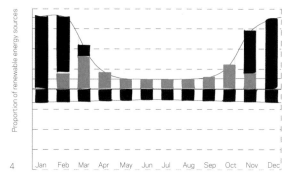

4   Jan   Feb   Mar   Apr   May   Jun   Jul   Aug   Sep   Oct   Nov   Dec

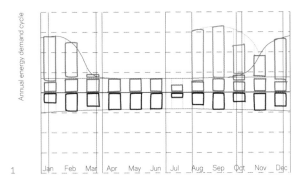

Annual energy demand cycle

1

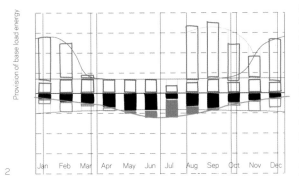

Provision of base load energy

2

Provision of peak load energy

3

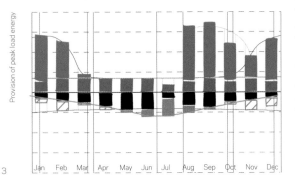

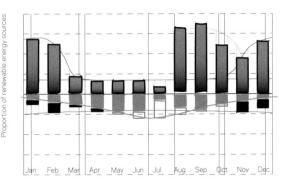

Proportion of renewable energy sources

4

SCHOOLS
Heating: district heating
Cooling: DEC system
Electricity: photovoltaics + mains electricity

1. Schools are a special case because of interruptions in use during the holiday periods lasting several weeks at different times of the year. , which directly affect the system design. The summer break significantly reduces the demand for cooling. No domestic hot water is required during summer unless the facilities, the gymnasium in particular, are used by others. During the winter break, the room temperature and mechanical ventilation are markedly reduced. Despite pauses in operation, an appreciable heating demand remains.

2. Owing to their high occupancy rates, schools are serviced by a central fresh air intake and extraction system. Therefore the considerable volumetric air flow of the mechanical ventilation is also used for heating/cooling the building.

3. Should the heat in summer cause uncomfortable conditions, it is customary for schools to close, but consideration could be given to utilizing a renewable source of cooling. Possible options are direct adiabatic cooling or a fresh air ground duct for preconditioning the intake air. The diagram analyses installation of DEC cooling integrated into the ventilation system. This is a type of sorption system based on thermally condensing the refrigerant, so the system requires high temperature heat rather than electricity for operation. Schools are often connected to a municipal district heating system for their heating supply. For this reason (and lacking demand in summer), solar thermal collectors are not installed. Instead, any suitable surfaces of the building can be used for the installation of photovoltaic systems.

4. A photovoltaic system can reduce the annual consumption of primary energy. However, as more electricity is produced in summer when the consumption is lower, much of the electricity produced is fed into the network. According to Paragraph 5 of EnEV 2009, no feed-in credits are received. The deciding parameter for the primary energy demand is the district heating. Depending on the source of the district heat and its form of generation, it may be derived entirely from renewable fuels or largely from fossil fuels.

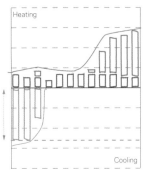

Heating

Cooling

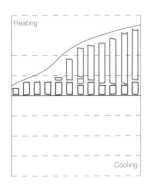

Heating

Cooling

Heating and cooling load diagrams.

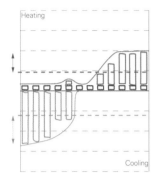

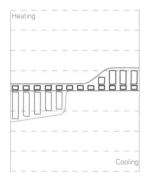

Heating and cooling load diagrams.

OFFICES (Option 1 of 3)
Heating: ground to water heat pump + condensing boiler
Cooling: ground to water heat pump + compression chiller
Electricity: photovoltaics + mains electricity

1. Modern office buildings have high internal sources of heat and hence a low demand for space heating; likewise the demand for domestic hot water heating is generally low. Therefore the supply of cooling and electricity for lighting, mechanical ventilation, and equipment are the dominating design factors. Nevertheless, there are substantial differences between base loads and peak loads in the supply of heating and cooling in the different seasons. This should be taken into account in the design of an energy concept by providing an intelligent combination of bifunctional generating and exchanging devices and by aiming to even out extremes of consumption.
2. Option 1 investigates utilization of a reversible ground-to-water heat pump with ground loop as a combined heating and cooling source. The ground loop accesses a relatively constant temperature level of around 10 °C regardless of the time of year. This can be used, for heating the building at a low flow temperature and for direct cooling. Activated building components are also used to support the heating/cooling. The heat pump cannot be used simultaneously for heating and cooling, so the system must be capable of reversing operation during the year. For the ground loop, it is important that the annual heat input and withdrawal are balanced. The constant demand for domestic hot water heating throughout the year should not be covered by the heat pump; if the heat pump were operated as B10/W50, its seasonal energy efficiency ratio would be significantly worse than as a B10/W35 system. Given the intention to balance the annual heat input and heat withdrawal from the ground, we arrive at the following situation: the base load for heating and cooling determines the size of the heat pump and the bore holes. The energy obtained from the ground loop is not "wasted" during the spring and autumn, but reserved for the extreme temperatures in winter and summer. 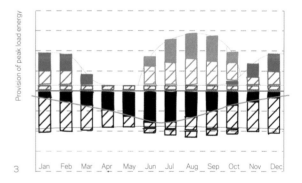 2
3. In this way, the remaining peak loads are reduced. The condensing boiler and the compression chiller can be smaller. All forms of generation are optimized in terms of utilization.
4. The demand for primary energy is significantly reduced by integrating renewable energy sources for part of the heating and cooling supply. The demand for electricity, critical from a primary energy standpoint, is partially covered by the photovoltaic installation; the residual demand is served by the mains network.

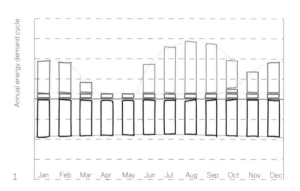

1   Jan   Feb   Mar   Apr   May   Jun   Jul   Aug   Sep   Oct   Nov   Dec

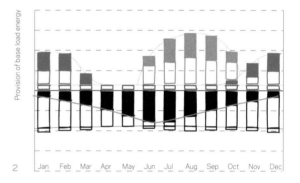

2   Jan   Feb   Mar   Apr   May   Jun   Jul   Aug   Sep   Oct   Nov   Dec

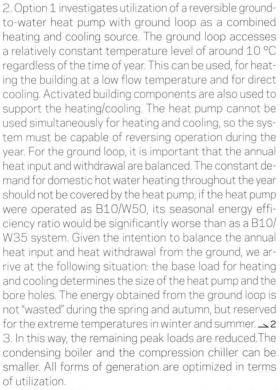

3   Jan   Feb   Mar   Apr   May   Jun   Jul   Aug   Sep   Oct   Nov   Dec

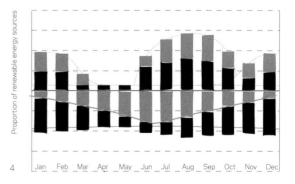

4   Jan   Feb   Mar   Apr   May   Jun   Jul   Aug   Sep   Oct   Nov   Dec

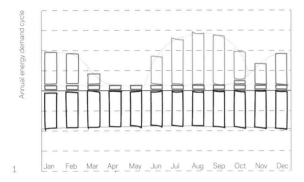

1

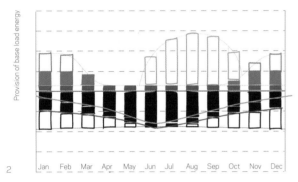

2

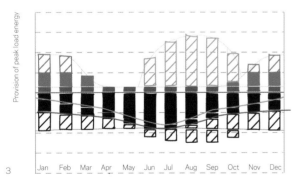

3

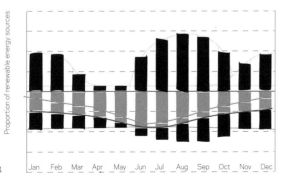

4

OFFICES (Option 2 of 3)
Heating: modular CHP (thermal) + condensing boiler
Cooling: compression chiller
Electricity: modular CHP (electrical) + photovoltaics + mains electricity

1. In Option 2 the energy concept is modified so that part of the demand for electricity is covered independently on site. In the first step we therefore investigate the installation of a combined heat and power unit. This consists of a gas powered generator which generates electricity and the waste heat from which provides part of the required heat. In order to achieve a maximum possible degree of utilization, the CHP unit is designed based on the heat produced. The production of electricity depends on the demand for heat—if this is low, the electricity generated on site will be less.

2. The CHP unit is designed to meet the base load of heat demand. This avoids oversizing the relatively expensive plant and avoids long periods of operation at partial load. The ratio of heat to electricity production is 65:35, meaning that a little more than half the quantity of heat generated is available as power for consumption in the building. This is technically referred to as an electrical output modulating range of 0.54. This is a good value for a Mini-CHP unit. Although the design concept does not rely on heat demand for domestic hot water, this steady base load benefits the design. This analysis is based on the assumption that domestic hot water is needed for staff showers.

3. Peak loads in this system are also covered by a condensing boiler using the same fuel as the CHP unit (natural gas or heating oil). The entire demand for cooling output is covered by an efficient compression chiller. It is possible to consider the installation of a heat pump as in the previous example; this could be powered from electricity generated by the CHP unit, but the installation of these relatively expensive plant components would lead to quite high investment costs. In addition, the CHP unit draws little benefit from this concept. A solar thermal installation would compete with utilization of the CHP unit and is thus excluded. As the office is in use practically throughout the day and the compression chiller is operated by electricity in summer, the proportion of electricity used from the PV system could be quite substantial. This simplifies load management and helps to create a favorable tariff structure.

4. This option can be further improved by replacing the fossil fuels used for the CHP unit and the cooling aggregate with renewable fuels.

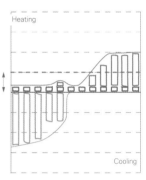

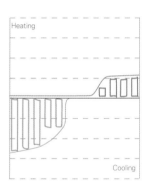

Heating and cooling load diagrams.

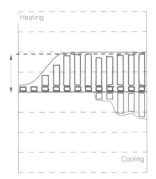

Heating and cooling load diagrams.

OFFICES (Option 3 of 3)
Heating: modular CHP (thermal)
Cooling: adsorption chiller + compression chiller
Electricity: modular CHP (electrical) + photovoltaics + mains electricity

1. The approach in Option 3 starts from the substantial differences of base and peak load heating and cooling in the different seasons. The aim is to achieve a leveling of demand and hence a more balanced supply. Suitable equipment for this purpose are bifunctional generating devices such as heat pumps and also sorption chillers, which we have also referred to as transformers. In the first step we therefore investigate an expansion of the CHP plant described in Option 2 to turn it into a combined cooling, heating, and power unit (CCHP). This involves an adsorption chiller which obtains its necessary heat input from the waste heat generated by the CHP unit. The diagram in the margin shows part of the cooling demand replaced by heat demand which is required for the sorption process.

2. The situation impacting on the design of the CHP unit is favorable: there is sufficient base load in winter and summer to ensure profitable operation. Only in spring, when no demand exists for heating or cooling, will the CHP unit be operated at partial load. If biogas is used for the CHP unit, the concept is particularly favorable vis-á-vis primary energy. Additionally, power generated by CHP earns, higher feed-in credits, compensating for the higher price of the biogas. Ideally, the savings in mains electricity costs are enough to pay for all the fuel used by the CHP unit. The heat produced by the CHP unit can be applied toward the depreciation of investment costs.

3. There is still the peak cooling load in summer to be covered by the relatively economical compression chiller. The electricity generated by the CHP and PV is still enough to satisfy the increased needs in the monthly balance. In summer and at weekends the excess electricity generated is fed into the mains network, which can be considered as compensation for insufficient energy production in spring and in the evenings.

4. This appears to make a fully renewables-based supply possible, but due to the high investment costs, the project must be analyzed for cost efficiency. The resulting combination of systems is rather complex and requires careful design and systematic coordination to supply a net zero energy office building with the greatest degree of comfort.

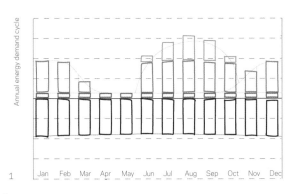

1

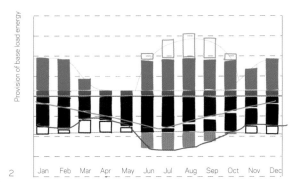

2

Geordnete Heiz- und Kühllast A3

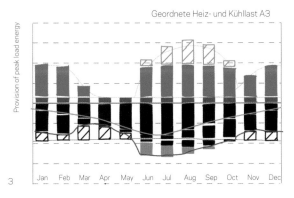

3

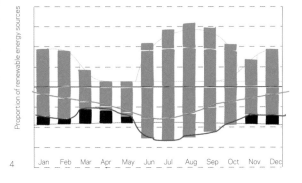

4

HEAT | COOL
CASE STUDIES

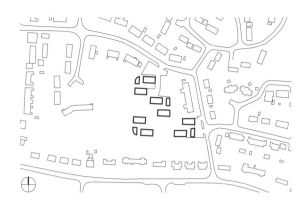

CLAY FIELD HOUSING
IN ELMSWELL, ENGLAND 2007
RICHES HAWLEY MIKHAIL ARCHITECTS, LONDON

Clay Field is the result of successful integrative cooperation between the building owner, the developer, the community and the government. The design of the project followed energy conservation principles, taking into account its rural location. The project comprises four residential units with six houses each, grouped around three communal gardens. Boundary walls between gardens are built of unfired clay blocks. A field pattern of swales (dips and hollows) is used to drain the open areas, which are reminiscent of agricultural areas. There are eight short terraces, each comprising three houses. The facades are clad with cedar shingles or stuccoed using a lime-based render. Variations in the facade design are achieved with different window sizes and positions. The windows were designed with the objective of creating an optimal ratio between passive solar gain and daylight. While the project was being developed, the residents' opinions were sought at each step, and the local primary school was involved in the project right from the beginning. Both the owners and the users benefited and increased their understanding by visiting the building site and engaging in lively discussions.

Site layout plan, scale 1:5000.

Staggered residential units with sloping roofs maximize solar gain.

Section, plans:
unit types a, b, c,
Partial southwestern layout with
outside area,
all 1:500 scale

The housing design picks up on
the rural character of Elmswell and
combines communal with private
open spaces.

a
1st floor   2nd floor

b
1st floor   2nd floor

c
1st floor   2nd floor   Attic

The houses are built in timber-frame construction with a sprayed-on insulation mixture consisting of lime and hemp. This construction guarantees low transmission and ventilation heat losses through its high degree of insulation and air-tightness. The floors are staggered and are linked by an open stairwell, which facilitates natural vertical ventilation. In winter, an additional mechanical ventilation system recovers 80% of heat from the exhaust air, which is used to heat the incoming air. In order to maximize solar gains, all houses are oriented towards the south. So as to avoid overshadowing adjacent buildings and plots as far as possible, the building units were designed with small cross sections and in varying sizes to allow the sun to fall deep into the buildings in winter. This also provides the buildings with the maximum possible level of daylight. For the generation of heat, a biomass boiler was installed, the fuel for which comes from locally available biomass material. In addition to passive strategies for reducing heating requirements, measures were also taken to conserve and recycle water.

The distinct, simple, and spatially exciting architecture has been developed from the regional building tradition (materials and color selection) and a clear concept for energy conservation.

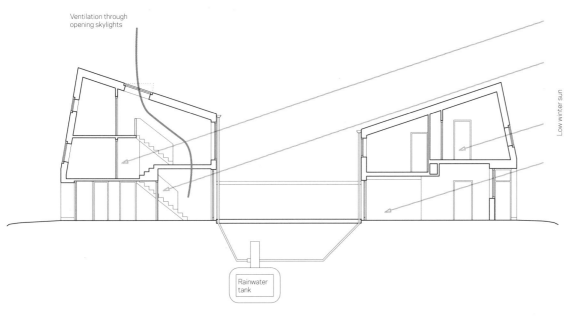

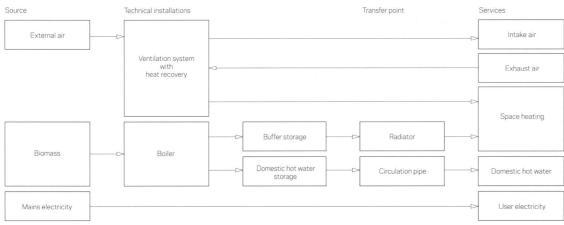

Resource concept, scale 1:2000.

System concept.

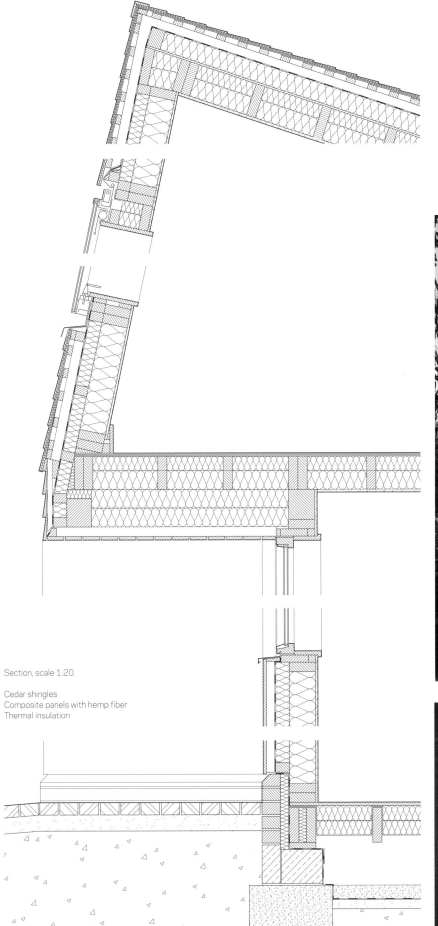

Section, scale 1:20.

Cedar shingles
Composite panels with hemp fiber
Thermal insulation

Kitchen with private
south-facing terrace.

Cedar shingle facade with entrance
loggias as seen from the footpath.

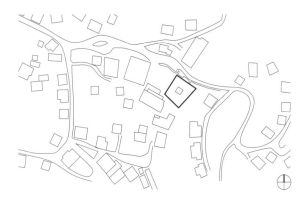

PASSIVHAUS PRIMARY SCHOOL
IN BRIXLEGG, AUSTRIA, 2007
RAIMUND RAINER, INNSBRUCK

The municipality of Brixlegg had its new primary school built according to Passivhaus principles in order to make a community contribution to climate protection. The building envelope has been insulated to optimum levels and the central ventilation system has been equipped with heat recovery so that conditions for teaching and learning are ideal at any time of the year. The heat requirement of the building is less than 15 kWh/m²a. Raimund Rainer and his office were able to achieve this high level of efficiency without expensive technical installations. Simple strategies achieve the same effect but cost less and are therefore better suited to the users. Entering from the schoolyard through the inset entrance, the visitor arrives at the large assembly hall, which can also be used for external events. There are large sliding doors which, in the case of fire, close the ground floor off from the heart of the school: an inner atrium which is lit from above, and around which the classrooms are grouped. Most corridors with their coat racks extend through to the facade in order to improve their spatial quality and create a relationship to the outside. Floors and ceilings in oak convey a feeling of value, and the fair-faced concrete and plastered walls provide thermal mass, which has a stabilizing effect on the interior climate at night.

Site layout plan, scale 1:5000.

Open space created between the secondary school, technical school, and primary school.

1st floor   2nd floor   Attic

Section, plans,
all 1:1000 scale.

Stairwell in central tower building.

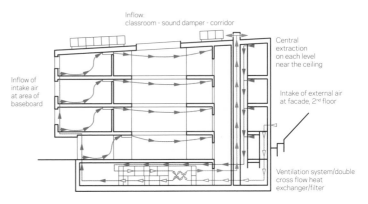

Tall windows and low sills improve daylight levels and the view. Automatically controlled external louvers protect against overheating. Intake air, with its temperature just below that of the room air, slowly enters the classroom through trickle air diffusers near the base of the facade. There are radiators to preheat the rooms; once classes have started, the internal heat sources are sufficient. As it warms up, the fresh air rises in the classroom and is conducted, through sound-insulated overflow openings beneath the ceiling, into the corridors, from where it is channeled through a duct to the central ventilation unit. In summer, a heat exchanger is used to extract un-wanted heat from the intake air into the exhaust air, which is first run through an adiabatic cooling process. In winter, the heat exchanger is used to reduce ventilation heat loss and to pre-warm the intake air, which then enters the classrooms through the trickle air diffusers at just the right temperature. A large part of the remaining heat requirement is covered by a 200 m^2 solar thermal installation. In the spirit of communal sharing of energy, any excess heat produced by the system is utilized for heating the water in the nearby open-air swimming pool. Any additional heat required is generated by an existing oil-fired boiler in the adjacent technical school.

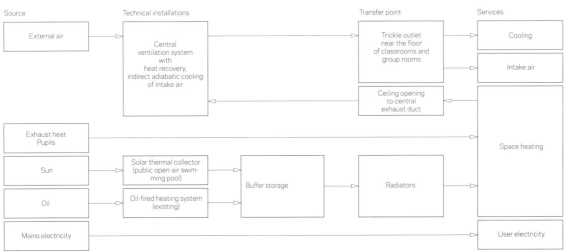

Classroom.

Ventilation concept, scale 1:500.

System concept.

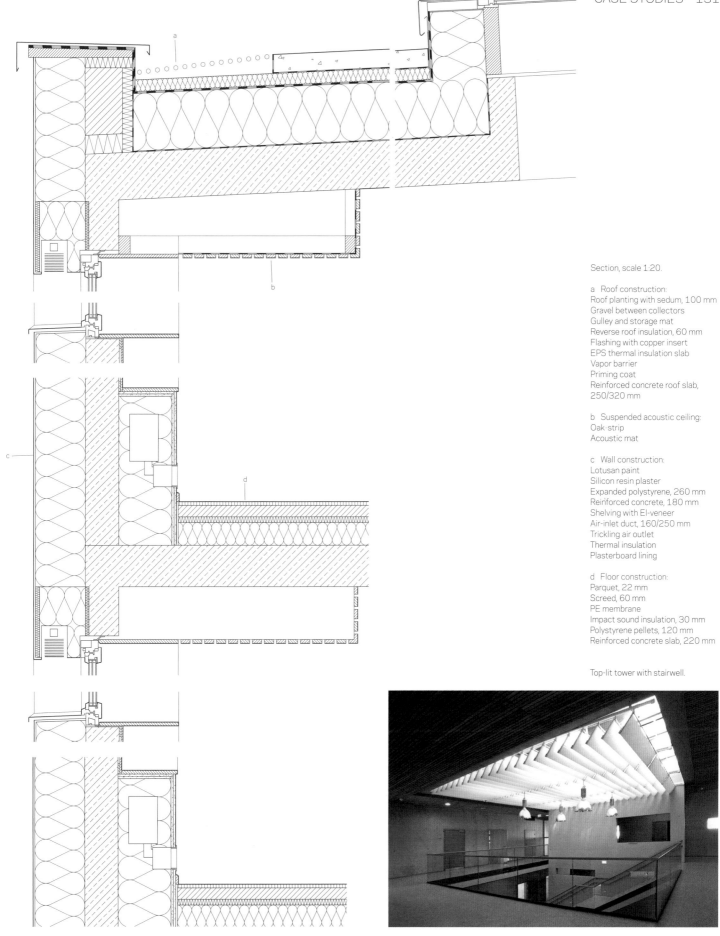

Section, scale 1:20.

a   Roof construction:
Roof planting with sedum, 100 mm
Gravel between collectors
Gulley and storage mat
Reverse roof insulation, 60 mm
Flashing with copper insert
EPS thermal insulation slab
Vapor barrier
Priming coat
Reinforced concrete roof slab,
250/320 mm

b   Suspended acoustic ceiling:
Oak-strip
Acoustic mat

c   Wall construction:
Lotusan paint
Silicon resin plaster
Expanded polystyrene, 260 mm
Reinforced concrete, 180 mm
Shelving with El-veneer
Air-inlet duct, 160/250 mm
Trickling air outlet
Thermal insulation
Plasterboard lining

d   Floor construction:
Parquet, 22 mm
Screed, 60 mm
PE membrane
Impact sound insulation, 30 mm
Polystyrene pellets, 120 mm
Reinforced concrete slab, 220 mm

Top-lit tower with stairwell.

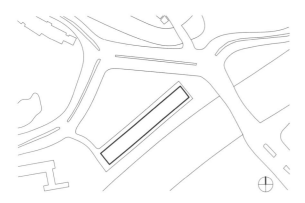

RIJKSWATERSTAAT ZEELAND HEAD OFFICE
IN MIDDELBURG, THE NETHERLANDS, 2004
ARCHITECTENBUREAU PAUL DE RUITER,
AMSTERDAM

The designer of the new administration building faced the task of pushing the existing town limit outwards and of developing virgin land. De Ruiter has given expression to this situation by designing a distinct and impressive building block. A location at the city boundary does not mean separation from the city. Here, the building was used to offer new vistas and transparency. The building forms part of a sequence of landmarks along a canal which previously contained the city area. It presents its elongated main facade to the city and, at the same time, profits from the views over the canal. The administra-

tion building, with its nearly 12,000 m² floor area, offers space for 450 staff, archives, a restaurant, conference rooms, and a fitness center. The building was primarily constructed in steel and concrete. Flexibility in the design of the structure, installations, and interiors allows for offices of varying sizes to reflect the needs of in-house and external staff members. All technical data and communication installations follow a facade grid pattern of 1.20 m. This makes it possible to connect the equipment required by users at any point.

Site layout plan, scale 1:5000.

The office building marks the transition between city and countryside.

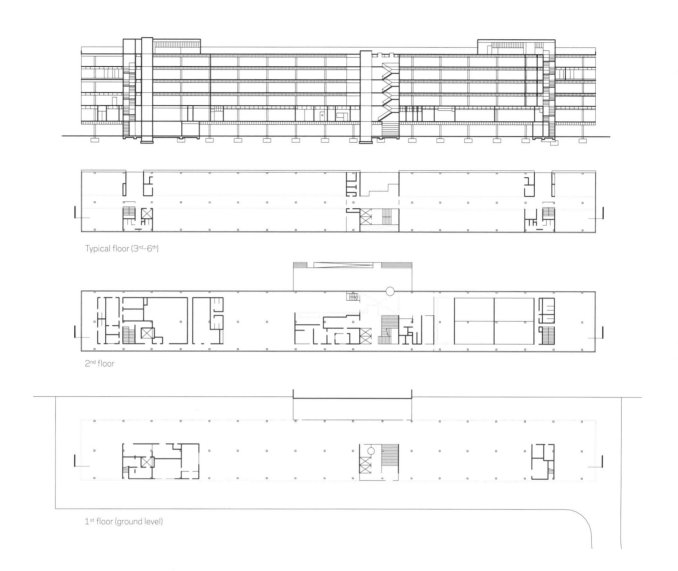

Typical floor (3rd–6th)

2nd floor

1st floor (ground level)

Section, plans,
all 1:1000 scale.

Open access areas as communication space.

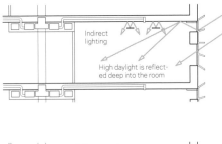

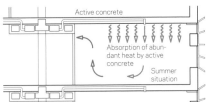

The transparent and clearly structured building is fully glazed towards the north-northwest, and semi-glazed towards the south-southeast. The high-quality glazing used in the climate facade combined with the excess heat generated by internal sources made it possible to allow generous glazing openings towards the north without unduly increasing the heating requirement. It also means that the electricity requirement for artificial lighting is relatively low, while overheating in summer is avoided due to the depth of the facade with its rigid projecting elements. There are no suspended ceilings in the building, which means that more daylight can enter the rooms at any time of day, with the amount controlled by a daylight control element at high level. The internal temperature is stabilized by the thermal mass of the exposed concrete components in the ceilings. These components are also utilized for conditioning the air of the building. Instead of heating the intake air with a dedicated system, this is achieved through activation of thermal mass. According to the simulation, this means that air changes could be reduced to one quarter of the amount otherwise required for cooling. Heating and cooling is provided by a ground-to-water heat pump. In this system, different layers of the ground are used for heat storage: depending on the groundwater situation, heating/cooling loops are cooled down and exhaust heat is "deposited" at different levels for the following winter.

The shape of the building is a systematic consequence of the energy concept used for office buildings. The utilization of internal heat gains and control of solar gains allow an open architecture without dominating the design.

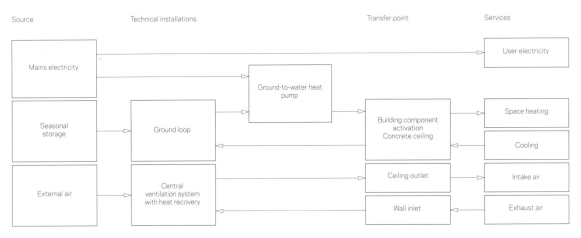

Ventilation and lighting concept, scale 1:200.

Corridor with translucent walls.

System concept.

Section, scale 1:20.

a   Wall construction:
Silver-gray fiber-cement panels,
8 mm
adhered to cover strips
Ventilated cavity 26 mm
Sheet-steel panel 1 mm
with rock-wool thermal insulation
140 mm

b   Solar screening louvers made of
aluminum

c   Internal Roman blinds

d   Double glazing in concealed
aluminum frame

e   Rock-wool thermal insulation
45 mm
Precast concrete balustrade
220 mm

f   Floor construction:
Flooring 3 mm
Anhydrite monolithic screed
50 mm
Lightweight poured concrete
160 mm
Precast concrete floor elements
100 mm

g   Openable floor hatch
Ventilation duct

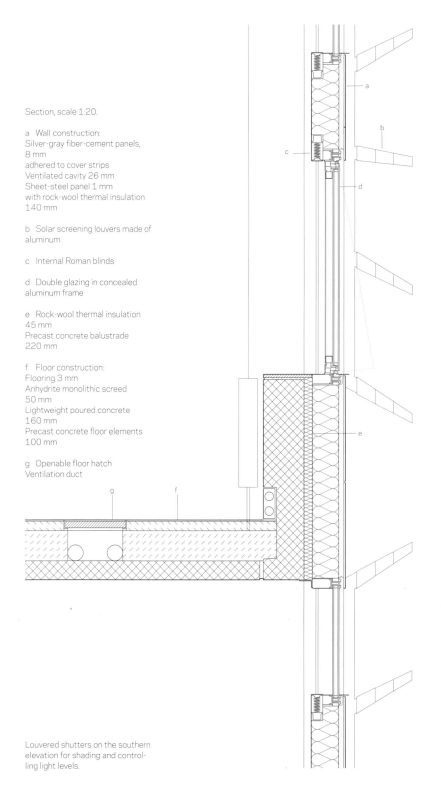

Louvered shutters on the southern
elevation for shading and controlling light levels.

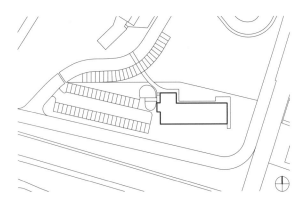

MARCHÉ INTERNATIONAL SUPPORT OFFICE
IN KEMPTTHAL, SWITZERLAND
KÄMPFEN FÜR ARCHITEKTUR AG, ZURICH

The freestanding building that serves as headquarters of the Marché restaurant chain has three stories accommodating fifty employees. The strict north-south orientation of the building has not been determined by the site or other buildings as the environment of a motorway service area provides hardly any points of reference. Thus it was possible to place the building so it derives maximum benefit from the sun. The southern elevation of the building features large glazed areas, providing solar gains in winter as well as good daylight. There are protruding balconies, which shade the windows in summer and thus prevent overheating. In order to minimize heat losses, the proportion of window area in the other facades has been reduced as much as possible. A photovoltaic installation completely covers the roof, which slopes towards the south to maximize yield. An unheated attic is larger towards the north side and serves as a buffer against cooling and overheating while reducing heat loss through the northern facade. Access to the flexible office space is via

a service core at the northwest of the building; the office space can thus be freely subdivided and furnished. The stairwells and the ground slab are the only reinforced concrete components of the building and have been constructed using recycled concrete. The remainder of the building is constructed using timber panels and it has an excellent overall eco-balance. The interior design is dominated by the warm surfaces of the untreated three-layer wall and ceiling panels. This impression is further enhanced by beech wood furniture designed specifically for the building. In addition, the wooden surfaces have been left untreated so they improve the interior room climate by absorbing excessive relative humidity and releasing it when there is demand. The high degree of prefabrication of the timber construction enabled the construction period to be significantly reduced. In addition, it was possible to integrate some elements of the services installations directly into the load-bearing structure, such as locating vertical ventilation ducts in the timber columns.

Site layout plan, scale 1:2000.

South elevation of building with balconies to protect against solar irradiation during summer.

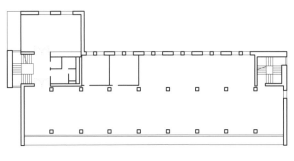

2nd floor

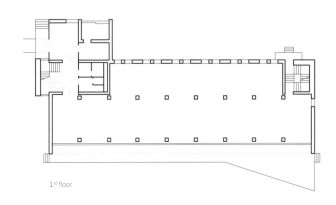

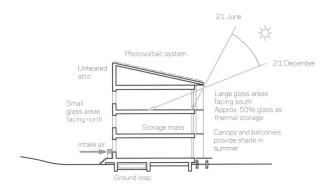

21 June

21 December

Photovoltaic system

Unheated attic

Small glass areas facing north

Large glass areas facing south Approx. 50% glass as thermal storage

Storage mass

Canopy and balconies provide shade in summer

Intake air

Ground loop

1st floor

Cross section showing the shadowing effect of the balconies and the orientation of the photovoltaic installation.

Plans, scale 1:500

Interior design with untreated wooden surfaces.

Stairwell core constructed using recycled concrete.

Roof covering with thin photovoltaics layer.

The rooms are heated using underfloor heating, which is supplied with hot water from a ground-to-water heat pump with two, 180 m long ground loops. In summer this can be used for passive cooling through the ground loop, although this is limited by our sense of comfort near floor level.

The whole building is equipped with a central ventilation system for exhaust and intake air. The external air is taken in through an earth duct beneath the ground slab, which means that it is preheated/pre-cooled, and blown into the rooms through openings near the floor. The rising warm air is extracted through openings at the ceiling. The whole system is controlled in each floor by CO_2 sensors. A vertical 12 m² hydroponics structure provides passive humidification of the room air. A pump delivers water to capillary matting with plants growing on it, which emit moisture to the room; any excess moisture is absorbed by the wooden surfaces.

The entire roof is covered by a 485 m² photovoltaic installation, which also serves as the roofing membrane. With an output of 44.6 kW_p and an annual electricity production of approximately 40,000 kWh, the system covers the electricity consumed by the services installations, and also that consumed by the users of the building.

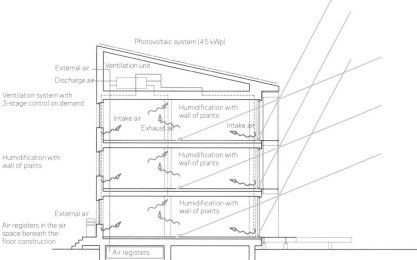

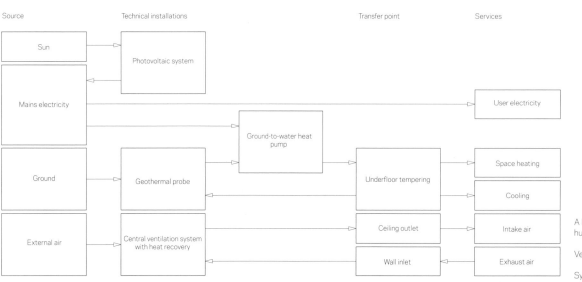

A hydroponics structure is used for humidifying the air in the offices.

Ventilation and lighting concept.

System concept.

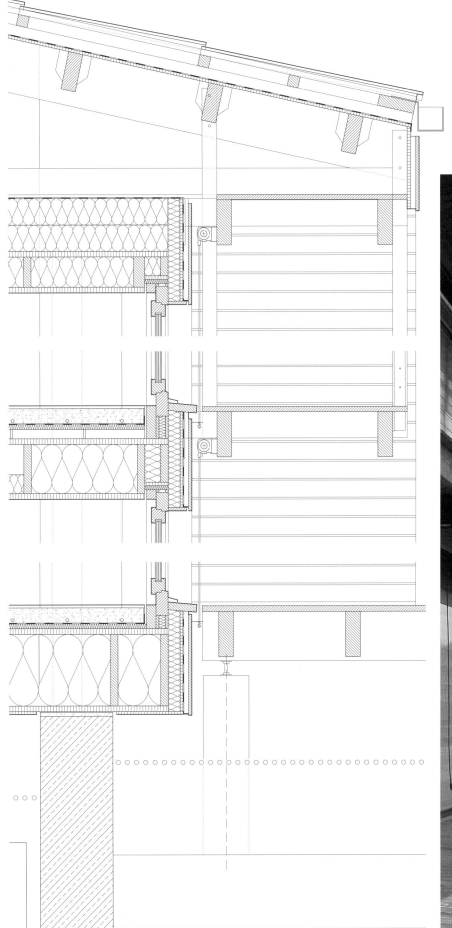

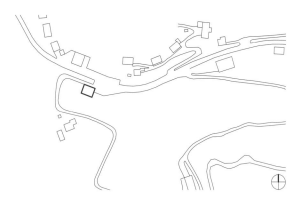

COMMUNITY CENTER
IN ST. GEROLD, AUSTRIA, 2009
CUKROWICZ NACHBAUR ARCHITECTS, BREGENZ

The new St. Gerold community centre in the Grosswalser Valley has created a popular focal point at the middle of this long, drawn-out village. It is part of a loose grouping of public buildings and, with its modest size, matches the scale of the other buildings in the valley. The compact freestanding building, which can be seen from afar, accommodates a number of functions on different levels: day nursery, play group, village shop, small multipurpose hall, and offices for the local administration. Owing to the building's location on sloping terrain, functions on the first floor, i.e. the day nursery, multipurpose hall and village shop, can still be accessed from level ground. The functions of the building have been arranged to suit the topography of the surrounding outside space. Towards the road there is a front yard with a fountain, while the day nursery is adjacent to a downward-sloping meadow. The layout provides circulation areas on the northern side of the building and accommodation rooms on the southern side. These two zones are separated by a narrow strip containing ancillary rooms and an elevator. The internal separating walls are not load-bearing and could be moved in order to allow alternative divisions. The building was constructed following Passivhaus principles, using energy efficient service installations; in addition, an ecological balance was conducted by making a life cycle assessment in order to minimize the effects on the environment. As a result, the entire building was constructed using timber elements of local silver fir and insulated with sheep's wool and wood fibers. Only the two lower floors, which are in contact with the ground, have been built of waterproof concrete. The construction timber and facade cladding, as well as all wood used for the interior, is left untreated, thus ensuring good air quality and an internal room climate free from noxious substances.

Site layout plan, scale 1:5000.

The building cube is subdivided by deep recessed access points and spacious openings.

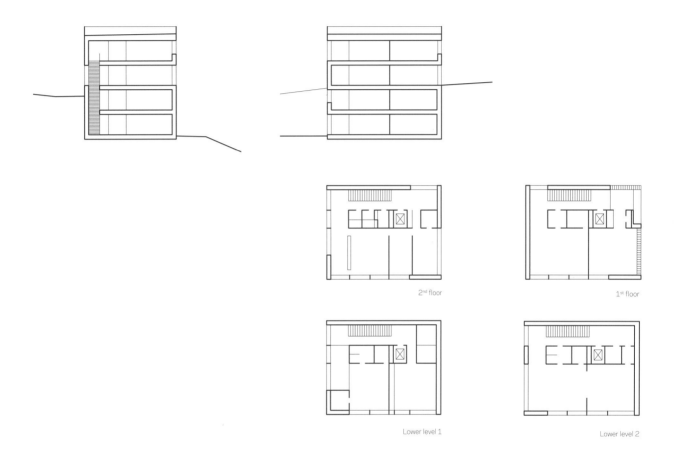

2nd floor

1st floor

Lower level 1

Lower level 2

Sections, plans,
all 1:500 scale.

View of entrance with forecourt.

For years, the St. Gerold community has been aiming for energy independence at a community level. In 2005 it was awarded the accolade "Energy Region of the Future." As part of their concept, buildings are individual components that contribute according to their potential. In spite of this commitment, installation of a photovoltaic system on the southern elevation was omitted for financial and design reasons, and also because a larger project was implemented nearby. The building impresses with its very low requirements for heat, cooling, and hot water heating. Space heating and cooling is provided by a ground-source heat pump, connected to two geothermal probes inserted to a depth of 160 m, which is capable of providing all the heat required. Over the course of the year this system achieves a coefficient of performance of 4.5 (B0/W35-COP 4.8). Heat is mainly transferred to the interior space by the hygienically required air change of 0.3 to 1.4 h^{-1}, which is controlled by CO_2 sensors. The efficiency of the circulating heat exchanger is 87%. Where the activities necessitate additional heating, intake air is heated by convectors. In order to minimize the energy impact of the cooling aggregates used in the village shop, the exhaust heat of the units is used for heating domestic hot water. Exterior solar screening louvers are automatically controlled with the help of a light sensor, which is also used for optimizing the use of daylight. A simulation study for the day nursery revealed room temperatures of just under 26 °C on a maximum of 13 days annually during summer, with the active cooling from the geothermal system switched off. All these days fall into the summer vacation period.

The timber construction of the building resulted from the life cycle analysis and determines the character of the interiors. Owing to the difference in the aging processes of the wood outdoors and indoors, the visual appearance of the building will be different internally and externally.

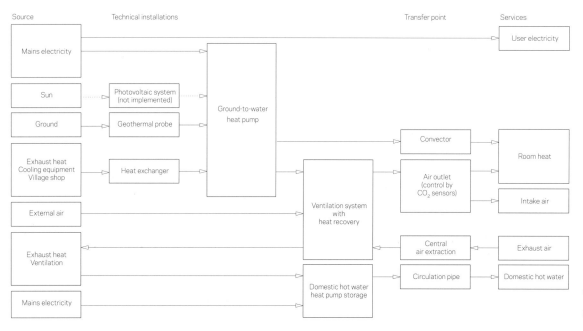

Source	Technical installations		Transfer point	Services

Mains electricity → User electricity

Mains electricity → Ground-to-water heat pump

Sun → Photovoltaic system (not implemented) → Ground-to-water heat pump

Ground → Geothermal probe → Ground-to-water heat pump

Exhaust heat Cooling equipment Village shop → Heat exchanger → Ground-to-water heat pump

External air → Ventilation system with heat recovery

Exhaust heat Ventilation

Mains electricity → Domestic hot water heat pump storage

Convector → Room heat

Air outlet (control by CO_2 sensors) → Room heat / Intake air

Central air extraction ← Exhaust air

Circulation pipe → Domestic hot water

Integrated shelves made of untreated solid wood.

System concept.

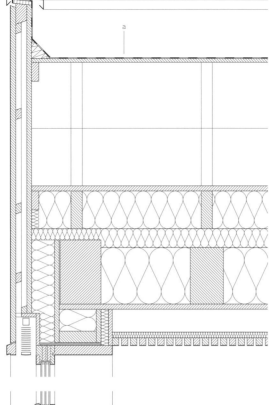

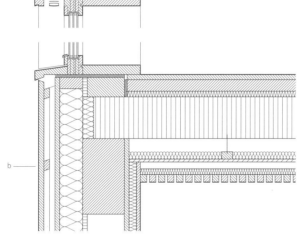

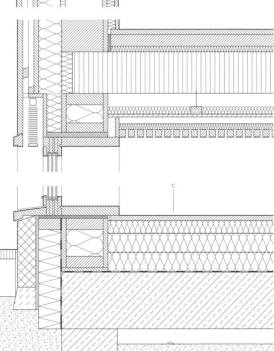

Section, scale 1:20.

a   Roof construction:
Bitumen roofing membrane,
with slate topping
Solid wood clapboard siding
(spruce)
Back ventilation, 500 mm
Solid wood clapboard siding
(spruce)
Timber battens, 180–300 mm,
with cellulose fiber insulation
Load-bearing timber deck,
300 mm,
with cellulose fiber insulation
Solid wood clapboard siding
(spruce)
Vapor check
Installation level
Acoustic insulation: sheep's wool,
30 mm
Dust membrane
Timber battens, 40 mm

b   Standard wall construction:
Face battens, silver fir, 30 mm
Base battens, 30 mm
Counter battens, 30 mm
Breather paper
Diagonal clapboarding (spruce),
25 mm
Timber battens, 125 mm
with cellulose fiber insulation
Diagonal clapboarding (spruce),
25 mm
Timber battens, 200 mm
with cellulose fiber insulation
Diagonal clapboarding (spruce),
25 mm
Vapor check
Timber battens, installation level
Silver fir clapboard

c   Floor construction:
Floorboards, 27 mm
Timber backing with insulation felt
Vapor barrier
Backing timbers
with wood fiber insulation, 100 mm
between
Backing timbers
with wood fiber insulation, 100 mm
between
Moisture barrier membrane
Priming coat
Reinforced concrete slab, 300 mm
Blinding layer, 80 mm

Upper level with citizen's service
provided by the municipal
administration.

Lower level with play group.

HEAT | COOL
APPENDIX

EXTERNAL FACTORS

Orientation	Pitch	Average monthly radiation intensity I_s in W/m²												Annual value kWh/m²a
		Jan	Feb	Mrz	Apr	Mai	Jun	Jul	Aug	Sep	Okt	Nov	Dez	Jan to Dec
Horizontal	0°	33	52	82	190	211	**256**	**255**	179	135	75	39	22	1120
South	30°	51	67	99	**210**	**213**	250	252	**186**	**157**	93	55	31	1216
	45°	57	**71**	**101**	205	200	231	235	178	**157**	**97**	59	34	1187
	60°	**60**	**71**	98	190	179	203	208	162	150	95	**60**	**35**	1104
	90°	56	61	80	137	119	130	135	112	115	81	54	33	810
Southeast/ Southwest	30°	45	62	**93**	203	211	248	251	183	149	87	49	28	1177
	45°	**49**	**64**	92	198	200	232	236	175	148	**88**	**51**	30	1142
	60°	**49**	62	88	185	182	208	213	161	140	85	**51**	30	1063
	90°	44	52	70	140	132	146	153	120	109	69	44	26	809
East/West	30°	**33**	**51**	**78**	181	199	238	240	170	129	72	**38**	21	1062
	45°	32	49	74	172	187	221	224	160	123	69	37	20	1002
	60°	30	46	68	160	171	201	205	148	114	65	35	19	923
	90°	25	37	53	125	131	150	156	115	90	51	28	15	713
Northwest/ Northeast	30°	**22**	**39**	**63**	151	180	222	221	150	105	57	**28**	16	918
	45°	20	35	56	132	158	194	194	133	91	51	26	14	808
	60°	18	32	49	116	139	168	170	118	81	46	23	13	711
	90°	14	25	38	89	105	124	128	90	62	35	18	10	541
North	30°	**20**	**34**	**54**	**137**	**173**	**217**	**214**	**142**	**90**	**49**	**26**	**15**	857
	45°	19	32	47	101	143	184	180	115	66	45	24	14	710
	60°	17	29	44	79	109	143	139	90	59	41	22	13	575
	90°	14	23	34	64	81	99	100	70	48	33	18	10	433
External air ϑ_e in °C		-1,3	0,6	4,1	9,5	12,9	15,7	18,0	**18,3**	14,4	9,1	4,7	1,3	8,9 °C
Days per month d_{mth} in d		31	28	31	30	31	30	31	31	30	31	30	31	365 d/a

DIN V 18599-10:2007-02, Table 7 – Values for radiation intensity and external temperatures for the reference climate in Germany

Average monthly radiation intensity on facades of different pitches and orientations with figures for irradiation as amount of energy for each month, and the annual total; location: Würzburg (reference used in the German DIN V 18599 standard). With reference to global radiation, a decrease can be observed from the south-east to the north-west (see map below).

Maximum radiation values are achieved in June and July. In July the maximum radiation intensities coincide with the highest external temperatures. Therefore the maximum cooling requirement is determined for the month of July. The figures indicate whether solar screening is required for a respective orientation. In addition, the function of the solar screening should also be checked for the month of September when the sun is lower.

The minimum weather-related radiation intensity of the sun for this geographical location is in December, while the lowest external temperatures are in January and February. A building's need for heating increases as temperatures fall and radiation intensities decrease. The maximum heat load or demand for domestic hot water of high insulated buildings is reached in January/February. Solar houses with spacious windows facing south reach the maximum mostly in December.

Average monthly radiation [W/m²] according to orientation and inclination

— Horizontal
---- South 30°
-- — South 60°
— South 90°
-·- East/West 90°
– – North 90°

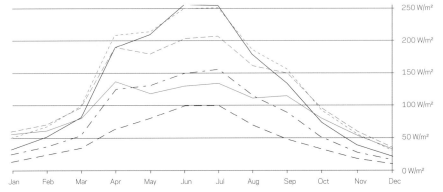

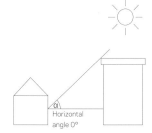

Horizontal angle 0°

The table shows the proportion of solar irradiation arriving at the surfaces of different orientations, in spite of shading by buildings in the vicinity. These so-called partial radiation factors affect thermal protection in summer (positive) as well as solar gains (negative).

Radiation factors for various surfaces depending on orientation, buildings in the vicinity and the season							
Horizontal angle	Season	North	NE/NW	East/West	SE/SW	South	Horizontal
0°	Winter	1.00	1.00	1.00	1.00	1.00	1.00
	Summer	1.00	1.00	1.00	1.00	1.00	1.00
10°	Winter	0.90	0.88	0.83	0.88	0.90	0.96
	Summer	0.88	0.88	0.91	0.94	0.96	0.98
20°	Winter	0.80	0.78	0.59	0.58	0.58	0.77
	Summer	0.80	0.74	0.79	0.86	0.93	0.90
30°	Winter	0.73	0.70	0.49	0.41	0.38	0.54
	Summer	0.75	0.63	0.65	0.76	0.88	0.76
40°	Winter	0.67	0.65	0.44	0.32	0.28	0.35
	Summer	0.71	0.55	0.53	0.64	0.78	0.57

DIN V 18599-2:2007-02. Table A.1 – Partial radiation factors F_h for various horizontal angles and pitches

U_g = U value of glazing only; this is different from the overall U value (U) which includes the Uf value for the frame
g_g = total energy transmittance of glazing
τ_e = radiation transmittance
τ_{d65} = light transmittance of glazing (visible part of radiation)

The light transmittance should be as high as possible, as it determines the daylight quality and intensity in the respective interior. By contrast, the total energy transmittance should be selected to suit the location as it affects the heat balance in the respective interior (compare thermal insulation glazing and solar screening glazing).

2S-WSV:
thermal insulation glazing (2 panes)
3S-WSV:
thermal insulation glazing (3 panes)
2S-SSV:
solar screening glazing (2 panes)

White awnings, foils, and roller blinds do not provide sufficient protection against glare. Fitting additional glare protection reduces light transmission but does not affect the g_{tot} value considerably.

Louver systems (blinds) are preferable in the 45° position. The values shown for a louver position of 10° have been calculated using the following weighting:
$g_{tot10°} = \frac{2}{3}\, g_{tot0°} + \frac{1}{3}\, g_{tot45°}$.

Energy-related performance data of glazing and solar screening devices and their combinations																		
Type of glazing	Performance data of glazing without solar screening				With external solar screening device						With internal solar screening device							
					Exterior blinds (10° position)		Exterior blinds (45° position)		Vertical awning		Interior blinds (10° position)		Interior blinds (45° position)		Roller blind		Foil
					white	dark gray	white	dark gray	white	gray	white	dark grey	white	dark grey	white	gray	white
	U_g	g_{tot}	τ_e	τ_{D65}	g_{tot}	g_{tot}	g_{tot}	g_{tot}	g_{tot}	g_{tot}	g_{tot}	g_{tot}	g_{tot}	g_{tot}	g_{tot}	g_{tot}	g_{tot}
Single	5.8	0.87	0.85	0.90	0.09	0.20	0.17	0.21	0.24	0.23	0.32	0.44	0.40	0.50	0.26	0.54	0.27
Couple	2.9	0.78	0.73	0.82	0.08	0.15	0.15	0.15	0.21	0.18	0.35	0.46	0.42	0.51	0.29	0.53	0.31
Triple	2.0	0.70	0.63	0.75	0.06	0.12	0.13	0.13	0.19	0.15	0.36	0.44	0.41	0.49	0.31	0.50	0.32
2S-WSV	1.7	0.72	0.60	0.74	0.06	0.11	0.12	0.11	0.19	0.14	0.36	0.45	0.42	0.50	0.31	0.52	0.32
2S-WSV	1.4	0.67	0.58	0.78	0.06	0.09	0.11	0.10	0.18	0.13	0.36	0.44	0.41	0.48	0.31	0.49	0.33
2S-WSV	1.2	0.65	0.54	0.78	0.05	0.08	0.11	0.09	0.17	0.12	0.36	0.44	0.41	0.47	0.31	0.49	0.33
3S-WSV	0.8	0.50	0.39	0.69	0.04	0.06	0.08	0.07	0.13	0.09	0.33	0.37	0.36	0.40	0.30	0.40	0.31
3S-WSV	0.6	0.50	0.39	0.69	0.03	0.05	0.08	0.05	0.13	0.08	0.33	0.38	0.36	0.40	0.30	0.40	0.31
2S-SSV	1.3	0.48	0.44	0.59	0.05	0.09	0.10	0.09	0.14	0.11	0.32	0.36	0.35	0.38	0.30	0.39	0.30
2S-SSV	1.2	0.37	0.34	0.67	0.04	0.08	0.08	0.09	0.12	0.10	0.27	0.30	0.29	0.31	0.26	0.31	0.26
2S-SSV	1.2	0.25	0.21	0.40	0.04	0.08	0.07	0.09	0.10	0.10	0.20	0.22	0.21	0.22	0.20	0.22	0.20
Performance data of solar screening device																	
Rate of transmission τ_{eB}					0.00	0.00	0.00	0.00	0.22	0.07	0.00	0.00	0.00	0.00	0.11	0.30	0.03
Rate of reflection ρ_{eB}					0.74	0.085	0.74	0.085	0.63	0.14	0.74	0.52	0.74	0.52	0.79	0.37	0.75

DIN V 18599-100:2009-10. Table 5 – Standard performance data

INTERNAL FACTORS

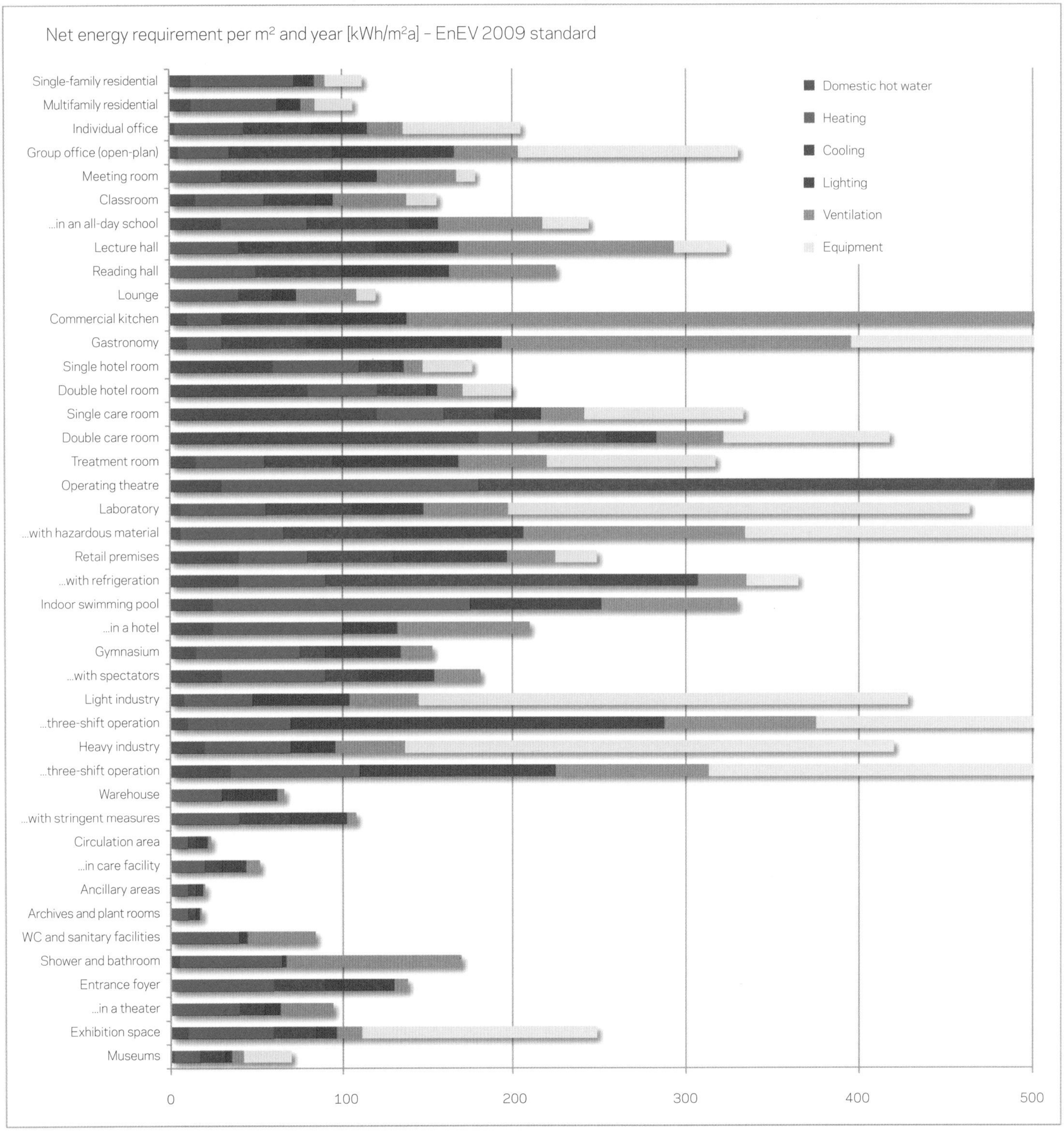
Net energy requirement per m² and year [kWh/m²a] – EnEV 2009 standard
Single-family residential
Multifamily residential
Individual office
Group office (open-plan)
Meeting room
Classroom
...in an all-day school
Lecture hall
Reading hall
Lounge
Commercial kitchen
Gastronomy
Single hotel room
Double hotel room
Single care room
Double care room
Treatment room
Operating theatre
Laboratory
...with hazardous material
Retail premises
...with refrigeration
Indoor swimming pool
...in a hotel
Gymnasium
...with spectators
Light industry
...three-shift operation
Heavy industry
...three-shift operation
Warehouse
...with stringent measures
Circulation area
...in care facility
Ancillary areas
Archives and plant rooms
WC and sanitary facilities
Shower and bathroom
Entrance foyer
...in a theater
Exhibition space
Museums
Domestic hot water
Heating
Cooling
Lighting
Ventilation
Equipment
0
100
200
300
400
500

Distribution of net energy requirements and primary energy requirements (see p. 150 f.) for different use zones under the new construction standards as defined in EnEV 2009.

The diagrams show the composition mix of area-related net energy requirements for heating, cooling and electricity per m² for a total of 42 use profiles. The net energy requirement for a whole building can be computed by adding the requirements for the different zones. The diagrams are based on the use profile overview tables with characteristic values for structure, category, and requirements on pages 152–155, as well as assumptions regarding the requirement for heating and cooling depending on the type of use. The distinction between EnEV 2009 and Passivhaus standards not only affects the heating requirement. The Passivhaus standard also covers higher efficiency lamps, fans, and electrical equipment, as these have an impact on the energy balance and there is a more exacting limit value.

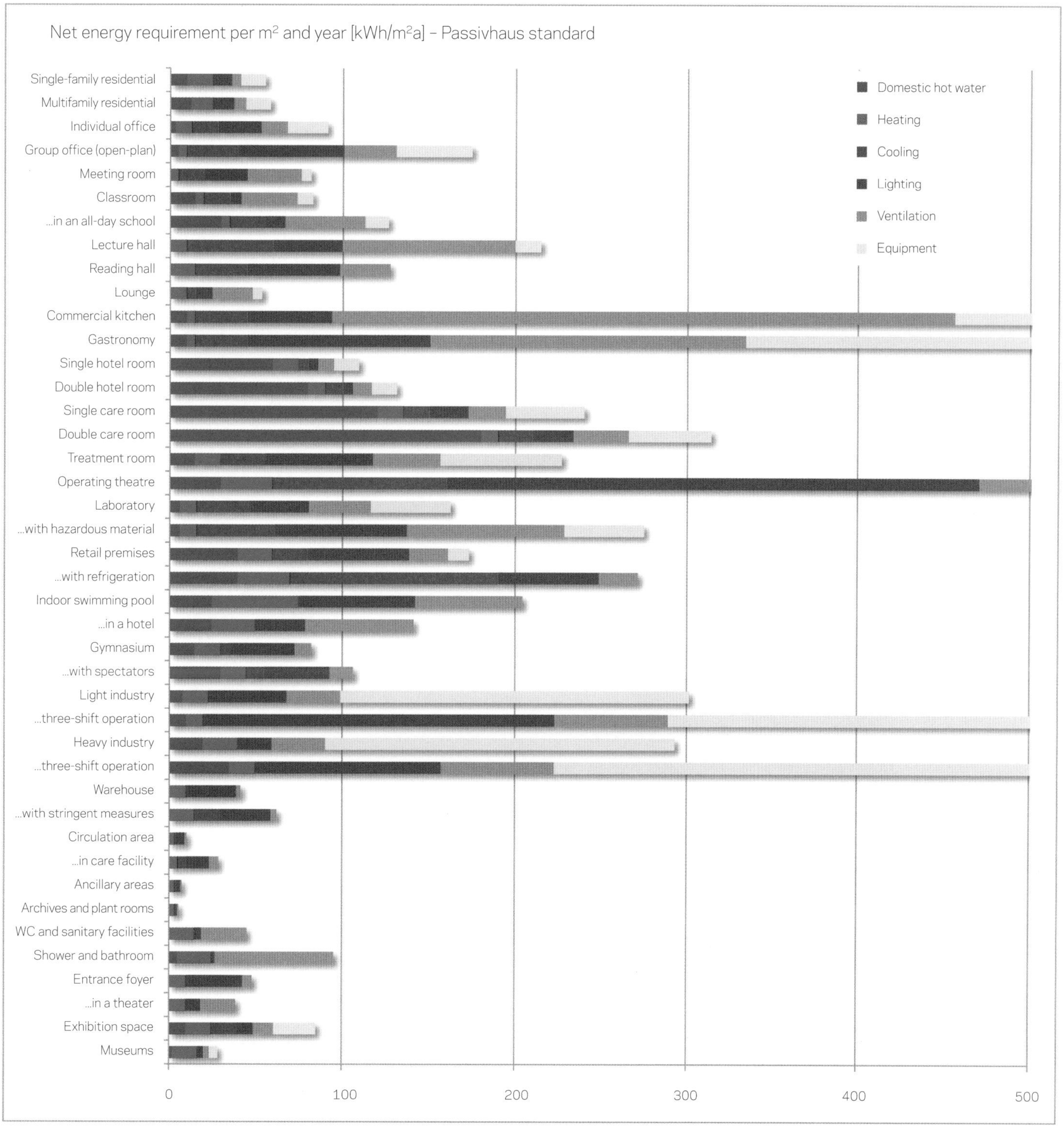
Net energy requirement per m² and year [kWh/m²a] – Passivhaus standard
Single-family residential
Multifamily residential
Individual office
Group office (open-plan)
Meeting room
Classroom
...in an all-day school
Lecture hall
Reading hall
Lounge
Commercial kitchen
Gastronomy
Single hotel room
Double hotel room
Single care room
Double care room
Treatment room
Operating theatre
Laboratory
...with hazardous material
Retail premises
...with refrigeration
Indoor swimming pool
...in a hotel
Gymnasium
...with spectators
Light industry
...three-shift operation
Heavy industry
...three-shift operation
Warehouse
...with stringent measures
Circulation area
...in care facility
Ancillary areas
Archives and plant rooms
WC and sanitary facilities
Shower and bathroom
Entrance foyer
...in a theater
Exhibition space
Museums
Domestic hot water
Heating
Cooling
Lighting
Ventilation
Equipment
0
100
200
300
400
500

Distribution of the net and primary energy requirement for different use zones for buildings built to a high energy conservation standard such as the Passivhaus standard.

The two diagrams are based on the distribution of net energy requirements, and are weighted according to primary energy. They show a typical mix of the primary energy requirement per m² for a total of 42 use profiles. Depending on the energy standard of the building, a primary energy factor per m² has been allocated to the respective services (domestic hot water, heating, cooling, electricity; see caption). This is derived from typical installation concepts: solar supported natural gas condensing boiler and compression cooling for new construction in accordance with EnEV 2009 (corresponds to the state-of-the-art technology of the EnEV reference building) and electrically operated ground-to-water heat pump for Passivhaus construction. The electricity requirement is assumed to be German mains current with a primary energy factor of currently 2.6.

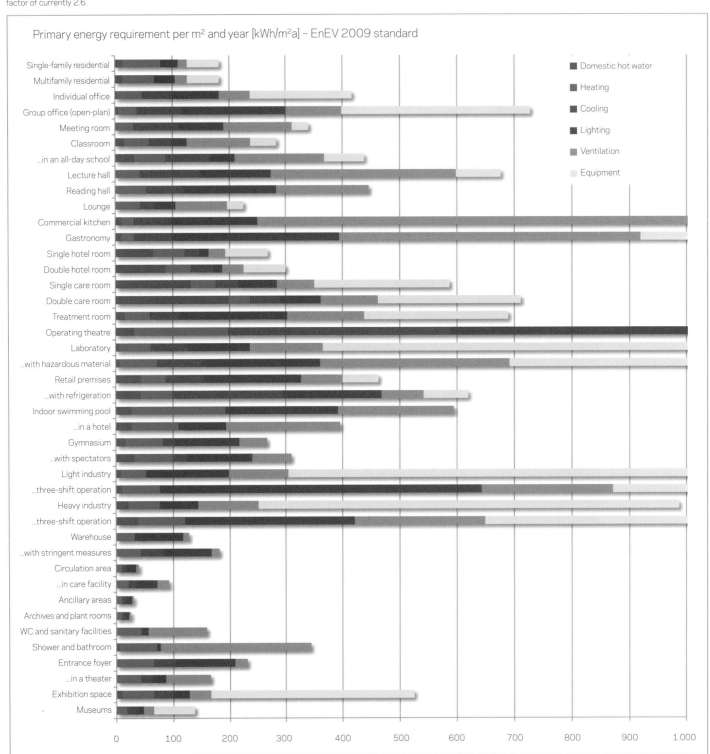

Primary energy requirement per m² and year [kWh/m²a] – EnEV 2009 standard

The following primary energy factors were used for the calculation: domestic hot water and heating fP = 1.1 (condensing boiler + solar thermal system); cooling fP = 1.3 (compression cooling); power input fP = 2.6 (mains electricity).

A comparison of the primary energy requirements shows the relevance of the electricity requirement, which becomes dominant in the case of certain uses and in buildings with high energy standards. This gives an indication of the quality of the heating and cooling technology available today. It is also evident that a further reduction in the primary energy requirement involves consideration of the electricity requirement for the building. A holistic approach should not stop at the definition of energy balance specified in EnEV. Even though the assessment of a daylight supply and the residual electrical requirement for lighting have been added to the energy balance following the introduction of DIN V 18599 in 2007, use-specific installation characteristics will not be evaluated under public law in the future, any more than they are now. However, this should not stop the services engineer from covering this area as part of his consultation.

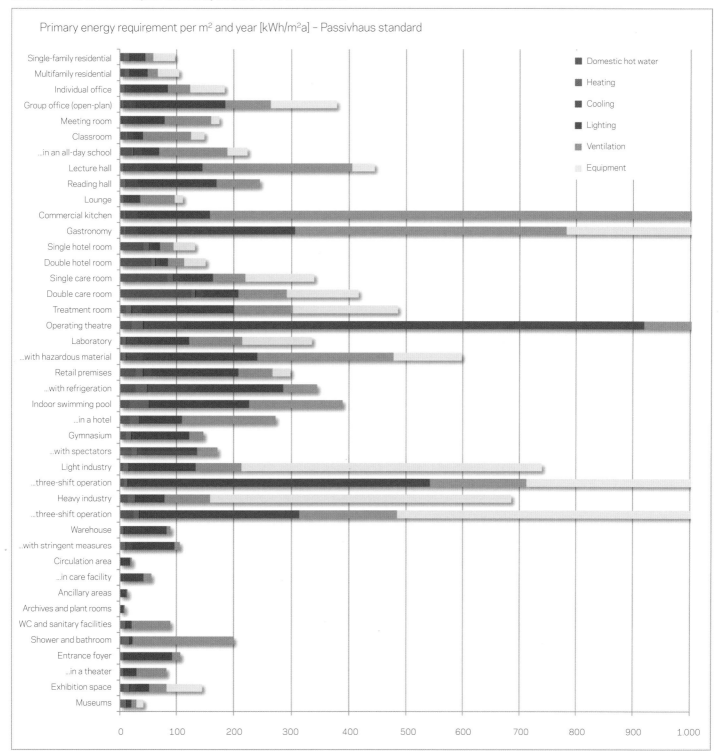

The following primary energy factors were used in the calculation: domestic hot water, heating and cooling fP = 0.7 (ground-to-water heat pump, COP (coefficient of performance) > 4.0); power input f_p = 2.6 (mains electricity).

Table 1

Use profile	Operating days per year	Operating time from/to	Hours	Operating hours day/night	Presence day/night	Hours of usage day/night	Floor area per person	Clear height	Room volume per person
	[d/a]	[time (24h clock)]	[h]	[h/a]	[%/d]	[h/d]	[m²/p]	[m]	[m³/p]
Single-family dwelling (multifamily dwelling)	365	0:00-24:00	24:00	4,407/4,353	40/80% 40/80%	1,763/3,482 1,763/3,482	35-45 m² 25-35 m²	2.5-3.0 m 2.4-3.4 m	88-135 m³ 60-119 m³
Office rooms for 1-6 desks (from 7)	250	7:00-18:00	11:00	2,543/207	70/70% 100/100%	1,780/145 2,543/207	10-18 m² 8-12 m²	2.5-4.0 m	25-72 m³ 20-48 m³
Meeting, seminar room	250	7:00-18:00	11:00	2,543/207	50/50%	1,272/104	2-4 m²	2.5-4.0 m	5-16 m³
Classroom, nursery school (all-day school)	200	8:00-14:00 8:00-17:00	6:00 9:00	1,200/0 1,700/100	75/75%	900/0 1,275/75	1.5-2.5 m²	2.5-4.0 m	4-10 m³
University lecture hall (library reading room)	150 300	8:00-18:00 8:00-20:00	10:00 12:00	1,409/91 2,999/601	75/75% 100/100%	1,057/68 2,999/451	0.8-1.2 m² 1.5-5.0 m²	4.0-8.0 m 3.0-5.0 m	3-10 m³ 5-25 m³
Other day rooms	250	7:00-18:00	11:00	2,543/207	50/50%	1,272/104	2-4 m²	2.5-4.0 m	5-16 m³
Kitchen for office canteen etc (commercial kitchen)	250 300	7:00-16:00 10:00-24:00	09:00 14:00	2,192/58 2,404/1,796	100/100%	2,192/58 2,404/1,796	10-12 m²	3.0-4.0 m	30-48 m³
Single bed hotel room (multibed room)	365	21:00-8:00	11:00	755/3,260	25/75%	189/2,445	8-12 m² 6-12 m²	2.5-3.0 m	20-36 m³ 15-30 m³
Bedroom, single (bedroom, multiple)	365	0:00-24:00	24:00	4,407/4,353	90/100% 100/100%	3,966/4,353 4,407/4,353	15-20 m² 10-14 m²	2.5-30 m	38-60 m³ 25-42 m³
Surgery, consulting room, treatment room (Operating theater)	250 365	7:00-18:00 7:00-22:00	11:00 15:00	2,543/207 3,936/1,539	80/80%	2,034/166 3,149/1,231	6-10 m² 5-8 m²	3.0-4.0 m 3.0-5.0 m	18-40 m³ 15-40 m³
Laboratory, few haz, subs, (many haz, subs,)	250	7:00-18:00	11:00	2,543/207	70/70%	1,780/145	10-18 m²	3.0-4.0 m	30-72 m³
Retail premises without refrigeration units (with)	300	8:00-20:00	12:00	2,999/601	100/100%	2,999/601	4-6 m²	3.0-8.0 m	12-48 m³
Swimming pool, indoor (in a hotel)	365	7:00-22:00	15:00	3,936/1,539	100/100% 50/50%	3,936/1,539 1,968/770	16-24 m²	8.0-15.0 m 3.0-4.0 m	128-360 m³ 48-96 m³
Gymnasium without spectators (with)	300	8:00-22:00	14:00	3,002/1,198	70/70%·	2,101/839	10-30 m² 5-15 m²	8.0-12.0 m	80-360 m³ 40-180 m³
Light industry (three-shift operation)	250 365	7:00-16:00 0:00-24:00	09:00 24:00	2,192/58 4,407/4,353	90/90%	1,973/52 3,966/3,918	15-25 m²	4.0-18.0 m	60-450 m³
Heavy industry (three-shift operation)	250 365	7:00-16:00 0:00-24:00	09:00 24:00	2,192/58 4,407/4,353	90/90%	1,973/52 3,966/3,918	15-25 m²	4.0-18.0 m	60-450 m³
Warehouse and logistics (high specification)	365	0:00-24:00	24:00	4,407/4,353	50/25%	2,204/1,088	40-80 m²	8.0-18.0 m	320-1440 m³
Circulation areas, (corridors, bedrooms)	250 365	7:00-18:00 0:00-24:00	11:00 24:00	2,543/207 4,407/4,353	20/20% 20/10%	509/41 881/435	10-20 m²	2.5-3.0 m	25-60 m³
Ancillary areas (archives, services rooms)	250	7:00-18:00	11:00	2,543/207	10/10% 5/5%	254/21 127/10	10-20 m²	2.5-3.0 m	25-60 m³
WC and sanitary facilities (showers and bathrooms)	250	7:00-18:00	11:00	2,543/207	10/10 % 5/5 %	254/21 127/10	10-20 m²	2.5-3.0 m	25-60 m³
Entrance foyer (Theater foyer)	250	7:00-18:00 19:00-23:00	11:00 04:00	2,543/207 55/954	75/75% 50/25%	1,907/155 28/236	10-20 m² 1-2 m²	3.0-8.0 m	30-160 m³ 3-16 m³
Exhibition rooms (conservation quality)	250	10:00-18:00	08:00	1,850/151	50/50% 10/10%	925/76 185/15	4-6 m² 10-30 m²	3.0-5.0 m 3.0-4.0 m	12-30 m³ 30-120 m³

Tables with overview of use profiles with characteristic values for structure, category, and requirements. Table 1 shows characteristic structural values. Although the different functions already determine different characteristics for the rooms in which they take place, they are here mainly used in order to be able to compare person-related data with those relating to space/structure.

The average operating days per year take into account the normal working week, public holidays, and holiday periods. The operating hours apply to daytime or nighttime (defined as hours after sunset and before sunrise) depending on the use function.

The *presence* column presents information about how frequently the effective use zone is in fact occupied by persons during the operating hours. Small rooms are unoccupied more frequently, whereas this is less rare with larger rooms. The actual usage hours are calculated from the operating periods and the hours of presence.

The requirement of square area per person is a statistical average. As with many other data, one needs to check whether the stated figures are plausible in the context of the actual use category. Where use of the room is more intensive, the respective figures will increase; conversely, where it is less intensive smaller values should be used.

The clear height refers to typical rooms in the respective use category, ranging from small rooms to large halls. The air volume per person is calculated from the floor area required per person and the height of the room.

Table 2

Heat Output Persons	Heat Output Rate Equipment	Heat Output Rate Lighting	Heat Output Rate Ventilation	Heat Output Rate Total	Heat Yield People	Heat Yield Equipment	Heat Yield Lighting	Heat Yield Ventilation	Heat Yield Total
[W/m²]	[W/m²]	[W/m²]	[W/m²]	[W/m²]	[Wh/m²d]	[Wh/m²d]	[Wh/m²d]	[Wh/m²d]	[Wh/m²d]
1,6-2,0 2,0-2,8	1-2	2.4-3.6	0.3-0.4 0.4-0.6	5.3-8.0 5.8-9.0	22.4-28.7 28.7-40.2	14.4-28.7 14.4-28.7	10.9-12.3 12.3-15.2	5.3-6.9 6.9-9.6	53-77 62-94
3.9-7.0 5.8-8.8	3-15 4-19	7.4-11.1 8.2-12.3	1.4-2.5 2.5-3.8	15.7-35.6 20.5-43.8	29.9-53.9 64.2-96.3	23.1-116 44.0-209	25.1-38 60.5-81	15.0-27.1 30.1-45.2	93-235 199-432
15.0-30.0	1-3	8.2-12.3	3.8-7.5	27.9-52.8	82.5-165	5.5-16.5	25.0-35.5	31.3-62.5	144-279
24.0-40.0	2-6	5.8-8.7	6.0-10.0 6.0-10.0	37.8-64.7	108-180 162-270	9.0-27.0 13.5-40.5	6.5-13.1 12.0-21.3	32.2-53.8 45.8-76.3	156-274 233-408
50.0-75.0 12.0-40.0	2-6 0-0	9.7-14.6 7.7-11.6	12.5-18.8 3.0-10.0	74.2-114 22.7-61.6	375-563 138-460	15.0-45.0 0.0-0.0	39.7-56.9 53.3-72.6	100-150 28.8-96.1	530-814 220-629
17.5-35.0	1-3	5.6-8.4	2.5-5.0	26.6-51.4	96.3-193	5.5-16.5	10.0-17.1	23.2-46.5	135-273
6.7-8.0	200-400 150-300	8.4-12.7	37.5-45.0 12.5-15.0	253-466 178-336	60.0-72.0 93.3-112	1800-3600 2100-4200	49-67 106-122	363-435 183-220	2271-4174 2482-4654
5.8-8.8 7.0-11.7	2-6	5.8-8.6	1.3-1.9 1.5-2.5	14.8-25.3 16.3-28.8	42.1-63.1 50.5-84.2	14.4-43.3 14.4-43.3	5.8-6.5 0.0-0.0	9.1-13.6 10.9-18.2	71-127 76-146
3.5-4.7 5.0-7.0	2-6	6.0-9.0	1.0-1.3 1.4-2.0	12.5-21.0 14.4-24.0	79.8-106 120-168	45.6-137 48.0-144	22.6-31 23.5-33	21.7-28.9 32.1-45.0	170-303 224-390
9.0-15.0 11.3-18.0	8-14 10-50	10.6-15.8 24.2-36.3	3.9-6.5 16.3-26.0	31.5-51.3 61.7-130	79.2-132 135-216	70.4-123 120-600	62.5-84 311-337	38.9-64.9 232-372	251-404 798-1525
5.0-9.0	6-36	8.1-12.1 9.7-14.6	3.5-6.3 9.0-26.3	22.6-90.4 29.7-102.8	38.5-69.3 38.5-69.3	46.2-485 46.2-485	34.7-49 76.2-85	35.5-63.9 91.1-164	155-667 252-803
11.7-17.5	1-3 -8-12	6.7-10.0	1.7-2.5	21.0-33.0 12.0-18.0	140-210 140-210	12.0-36.0 -96.0--144	58.6-75.3 0.0-0.0	22.5-33.8 22.5-33.8	233-355 67-100
2.9-4.4	0-0	7.2-10.8 4.8-7.2	3.7-5.5	13.8-20.7 11.4-17.1	43.8-65.6 21.9-32.8	0.0-0.0 0.0-0.0	67.1-86.5 28.8-35.3	62.3-93.5 62.3-93.5	173-246 113-162
3.7-11.0 6.0-18.0	0-0	7.2-10.8	1.0-3.0 1.3-4.0	11.9-24.8 14.5-32.8	35.9-108 58.8-176	0.0-0.0 0.0-0.0	37.8-50.4 0.0-0.0	9.5-28.4 13.5-40.5	83-187 72-217
3.2-5.3	25-45	10.8-16.2	3.0-5.0	42.0-71.5	25.9-43.2 69.1-115	203-365 540-972	45-67 203-232	30.7-51.1 65.8-109.6	304-525 878-1429
4.0-6.7	25-45	6.1-9.2	3.0-5.0	38.1-65.8	32.4-54.0 86.4-144	203-365 540-972	20-32 107-124	30.7-51.1 65.8-109.6	285-501 799-1349
1.1-2.3	0-0	4.3-6.5	0.2-0.4	5.6-9.1	10.1-20.3 10.1-20.3	0.0-0.0 0.0-0.0	29.1-35.7 0.0-0.0	2.6-5.3 3.6-7.1	42-61 14-27
3.5-7.0	0-0	2.8-4.2 3.5-5.3	0.1-0.3 0.8-1.5	6.4-11.5 7.8-13.8	7.7-15.4 12.6-25.3	0-0 0-0	6-7 14-15	1.2-2.4 5.3-10.5	15-25 32-50
3.5-7.0	0-0	2.8-4.2	0.1-0.1	6.4-11.3	3.9-7.7 1.9-3.9	0.0-0.0 0.0-0.0	3.9-3.9 1.9-1.9	0.7-1.4 0.7-1.4	8-13 5-7
3.5-7.0	0-0	4.8-7.2	1.9-3.8 4.9-9.8	10.2-18.0	3.9-7.7 1.9-3.9	0.0-0.0 0.0-0.0	4.2-5.4 2.1-2.7	26.3-52.5 68.3-137	34-66 72-143
3.5-7.0 35.0-70.0	0-0	7.2-10.8	0.5-1.0 5.0-10.0	11.2-18.8 47.2-90.8	28.9-57.8 36.9-73.9	0.0-0.0 0.0-0.0	33.1-46.8 8.9-9.1	5.6-11.1 20.1-40.3	67-116 66-123
15.0-22.5 3.0-9.0	6-63	4.2-6.3 4.2-6.3	1.7-2.5 0.4-1.3	26.9-94.3 13.7-79.6	60.0-90.0 2.4-7.2	24.0-252 4.8-50.4	9.4-13 3.4-3.8	11.7-17.5 3.5-10.4	105-373 14-72

Table 2 shows the *heat output rate* and *heat yield* for different sources. The heat output rate is the heat given off by people, equipment, lighting and mechanical ventilation in operation per m². The heat yield, which is the daily average heat produced, is calculated by multiplying the respective heat output rate by the hours a person is in the room and by the operating hours of equipment respectively.

The heat produced by people depends on their activities and varies between 60 and 120 W. The values entered are only those for "dry" heat, without moisture content. Other factors are density (people/m²) and presence. The heat input from people is the only value that does not result in an energy requirement and is therefore independent of a technological solution, but instead depends on human behavior.

Data for the heat produced by equipment such as computers or other machines have mostly been taken from DIN V 18599. The energy required for the use of equipment can vary greatly between users; it depends on technical factors (type and number of items) as well as on user behavior (use efficiency).

The heat produced by artificial lighting is calculated from the light output rate, which in turn depends on the level of lighting, the type of light fitting, and user behavior. The figures are based on efficient light sources such as fluorescent strip lighting and LED light fittings. While the quality of the daylight provision per square meter does not impact on the heat output rate, it heavily affects the resulting heat yield.

The heat produced by mechanical ventilation is the heat given off by fans. The data are based on air extraction and fresh air intake with heat recovery and filter classes typical for the respective use. Heat gains from differences in the temperature of intake and exhaust air vary according to equipment and season and therefore have to be considered separately for the respective building and location, similar to the other, external, factors (solar irradiation, heat transmission through the building envelope, etc.).

Table 3

Use profile	Illumination level, target	Specific power output from lighting	Degree of cover through natural lighting	Artificial lighting operating hrs.	Electricity requirement for lighting	Electricity requirement for lighting
	[lx]	[W/m²]	[%]	[h/d]	[Wh/m²d]	[kWh/m²a]
Single-family dwelling (multifamily dwelling)	200	3.0	90-80% 80-60%	3.62-4.11 4.11-5.07	10.9-12.3 12.3-15.2	4.0-4.5 4.5-5.6
Office rooms for 1-6 desks. (from 7)	500	9.2 10.2	70-50% 50-30%	2.72-4.14 5.91-7.95	25.1-38.3 60.5-81.3	6.3-9.6 15.1-20.3
Meeting. seminar room	500	10.2	60-40%	2.45-3.47	25.0-35.5	6.3-8.9
Classroom. nursery school (all-day school)	300	7.3	80-60%	0.90-1.80 1.65-2.93	6.5-13.1 12.0-21.3	1.3-2.6 2.4-4.3
University lecture hall (library reading room)	500	12.1 9.7	60-40%	3.72-4.68 5.50-7.50	39.7-56.9 53.3-72.6	6.0-8.5 8.0-10.9
Other day rooms	300	7.0	80-60%	1.43-2.45	10.0-17.1	2.5-4.3
Kitchen for office canteen etc. (commercial kitchen)	500	10.6	50-30%	4.62-6.37 9.99-11.60	48.7-67.3 105.5-122.5	12.2-16.8 31.7-36.7
Single bed hotel room (multi-bed room)	200	7.2	80-60%	0.80-0.91	5.8-6.5	2.1-2.4
Bedroom. single (bedroom. multiple)	300	7.5	90-80%	3.01-4.10 1.13-4.34	22.6-30.7 23.5-32.6	8.2-11.2 8.6-11.9
Surgery. consulting room. treatment room	500 1000	13.2 30.2	50-30% 20-10%	4.73-6.36 10.27-11.14	62.5-83.9 310.7-336.8	15.6-21.0 113.4-112.9
Laboratory. few hazardous substances (many haz. subs.)	500	10.1 12.1	60-40% 20-10%	3.43-4.85 6.28-6.99	34.7-49.1 76.2-84.9	8.7-12.3 19.1-21.2
Retail premises without refrigeration units (with)	300	8.4	50-30%	7.00-9.00	58.6-75.3	17.6-22.6
Swimming pool. indoor (in a hotel)	300 200	9.0 6.0	70-50% 50-30%	7.45-9.61 4.80-5.88	67.1-86.5 28.8-35.3	24.5-31.6 10.5-12.9
Gymnasium without spectators (with)	300	9.0	80-60%	4.20-5.60	37.8-50.4	11.3-15.1
Light industry (three-shift operation)	500	13.5	60-40%	3.37-4.94 15.08-17.25	45.3-66.6 203.0-232.3	11.3-16.6 50.8-58.1
Heavy industry (three-shift operation)	300	7.7	70-50%	2.58-4.15 13.99-16.17	19.7-31.8 107.0-123.7	4.9-7.9 26.8-30.9
Warehouse and logistics (high specification)	150	5.4	60-40%	5.40-6.60	29.1-35.7	10.6-13.0
Circulation areas. (corridors. bedrooms)	100 125	3.5 4.4	20-10%	1.79-2.00 3.12-3.37	6.3-7.0 13.7-14.7	1.6-1.7 3.4-3.7
Ancillary areas (archives. services rooms)	100	3.5	0-0%	1.10-1.10 0.55-0.55	3.9-3.9 1.9-1.9	1.0-1.0 0.5-0.5
WC and sanitary facilities (showers and bathrooms)	200	6.0	40-20%	0.69-0.90 0.35-0.45	4.2-5.4 2.1-2.7	1.0-1.3 0.5-0.7
Entrance foyer (theater foyer)	300	9.0	60-40%	3.67-5.20 0.99-1.01	33.1-46.8 8.9-9.1	8.3-11.7 2.2-2.3
Exhibition rooms (conservation quality)	200	5.3 5.3	60-40% 20-10%	1.78-2.52 0.65-0.73	9.4-13.3 3.4-3.8	2.4-3.3 0.9-1.0

Table 3 shows the specific characteristics for artificial lighting. The listed characteristics in Tables 3 and 4, operating hours and operating modes are based on the structural data from Table 1 and form the basis for the resulting heat output and heat input shown in Table 2.

The target illumination level has been taken from the relevant standards. The specific lamp output assumes an efficient light source and mainly direct illumination. The proportion of natural lighting is an estimated value which takes into account the typical room depth, the position of the zone within the building, and how much rooms are used during daytime and nighttime. Finally, the electricity requirement results from multiplying the lamp output by the artificial lighting operating hours. The estimate of operating hours assumes that artificial lighting is used only when needed, i.e. it is switched on and off manually or by means of automatic presence and daylight sensors.

Table 4

Operating days of mechanical ventilation	Operating hours full/part/base/average				Flow volume full/part/base/average				Air changes daily average	Flow volume average load	Flow volume peak load	Specific electricity requirement exhaust/intake air/heat recovery/ humidity recovery/filtering					Electricity requirement exhaust/intake air/heat recovery/ humidity recovery/filtering				
[d/a]	[h/d]				[m³/h P]				[1/h]	[m³/m²h]	[m³/m²h]	[Wh/m²d]					[kWh/m²a]				
250	9.6	4.8	9.6	16.0	30	20	10	19.98	0.23-0.15	0.44-0.57	0.67-0.86	2.4	3.0	0.6	0.6	1.2	0.61	0.76	0.15	0.15	0.30
	10	5	10	16.0	30	20	10	19.98	0.33-0.17	0.57-0.80	0.86-1.20	3.3	4.1	0.8	0.8	1.6	0.82	1.03	0.21	0.21	0.41
200	8	3	3	10.8	50	30	20	38.64	1.55-0.54	2.15-3.86	2.78-5.00	8.4	10.5	2.1	2.1	4.2	1.68	2.10	0.42	0.42	0.84
	10	4	0	12.0	60	30	20	51.61	2.58-1.08	4.30-6.45	5.00-7.50	15.1	18.8	3.8	3.8	7.5	3.01	3.76	0.75	0.75	1.51
250	6	3	6	8.3	30	10	10	17.86	3.57-1.12	4.46-8.93	7.50-15.00	18.8	23.4	4.7	4.7	9.4	4.7	5.9	1.2	1.2	2.3
200	4	2	2	5.4	30	10	10	21.5	5.73-2.15	8.60-14.33	12.00-20.00	17.2	21.5	4.3	4.3	8.6	3.44	4.30	0.86	0.86	1.72
	6	2	2	7.6	30	10	10	21.79	5.81-2.18	8.71-14.52	12.00-20.00	24.4	30.5	6.1	6.1	12.2	4.88	6.10	1.22	1.22	2.44
150	6	3	3	8.0	30	10	10	20.87	6.52-2.17	17.39-26.09	25.00-37.50	50.0	62.5	12.5	12.5	25.0	7.50	9.38	1.88	1.88	3.75
300	8	5	1	9.6	30	10	10	21.36	4.75-0.85	4.27-14.24	6.00-20.00	25.0	31.2	6.2	6.2	12.5	7.50	9.37	1.87	1.87	3.75
200	5	4	6	9.3	20	10	10	13.27	2.65-0.83	3.32-6.64	5.00-10.00	13.9	17.4	3.5	3.5	7.0	2.8	3.5	0.7	0.7	1.4
250	9	2	0	9.7	900	300	20	790.9	26.36-16.48	65.91-79.09	75.00-90.00	159.5	199.4	39.9	39.9	79.8	39.88	49.84	9.97	9.97	19.94
300	14	2	0	14.7	300	100	20	275	9.17-5.73	22.92-27.50	25.00-30.00	80.7	100.8	20.2	20.2	40.3	24.20	30.25	6.05	6.05	12.10
250	4	4	4	7.3	30	20	10	19.84	0.99-0.55	1.65-2.84	2.50-3.75	4.5	5.7	1.1	1.1	2.3	1.14	1.42	0.28	0.28	0.57
	4	4	4	7.3	30	10	10	19.84	1.32-0.66	1.98-3.31	3.00-5.00	5.8	7.3	1.5	1.5	2.9	1.45	1.82	0.36	0.36	0.73
250	17	6	1	21.7	40	30	10	36.12	0.96-0.60	1.81-2.41	2.00-2.67	10.1	12.6	2.5	2.5	5.1	2.53	3.16	0.63	0.63	1.26
	18	6	0	22.5	40	30	10	37.5	1.50-0.89	2.68-3.75	2.86-4.00	15.4	19.3	3.9	3.9	7.7	3.86	4.82	0.96	0.96	1.93
250	8	3	2	10.0	60	30	20	46.08	2.56-1.15	4.61-7.68	6.00-10.00	16.0	20.0	4.0	4.0	8.0	3.99	4.99	1.00	1.00	2.00
365	10	4	3	14.3	200	200	20	168.2	11.22-4.21	21.03-33.65	25.00-40.00	93.0	116.2	23.2	23.2	46.5	33.93	42.41	8.48	8.48	16.96
200	8	2	3	10.2	125	125	20	98.35	3.28-1.37	5.46-9.83	6.94-12.50	19.9	24.9	5.0	5.0	9.9	3.98	4.97	0.99	0.99	1.99
250	8	2	3	10.1	250	250	30	194.2	6.47-2.70	10.79-19.42	13.89-25.00	39.3	49.1	9.8	9.8	19.6	9.82	12.27	2.45	2.45	4.91
300	12	3	0	13.5	20	10	10	18	1.50-0.38	3.00-4.50	3.33-5.00	11.3	14.1	2.8	2.8	5.6	3.4	4.2	0.8	0.8	1.7
300	12	3	0	13.5	20	10	10	18	1.50-0.38												
365	15	2	0	17.0	160	160	160	160	1.25-0.44	6.67-10.00	6.67-10.00	28.3	35.4	7.1	7.1	14.2	10.3	12.9	2.6	2.6	5.2
	8	2	8	17.0	160	160	160	160	1.25-0.44												
200	8	4	4	9.5	60	10	10	35.52	0.44-0.10	1.18-3.55	2.00-6.00	7.6	9.5	1.9	1.9	3.8	1.52	1.89	0.38	0.38	0.76
	8	4	4	10.1	40	10	10	25.31	0.63-0.14	1.69-5.06	2.67-8.00	10.8	13.5	2.7	2.7	5.4	2.16	2.70	0.54	0.54	1.08
150	8	2	1	10.2	150	150	20	139.4	2.32-0.31	5.57-9.29	6.00-10.00	16.4	20.4	4.1	4.1	8.2	2.45	3.07	0.61	0.61	1.23
200	22	0	2	21.9	150	150	20	137	2.28-0.30	5.48-9.13	6.00-10.00	35.1	43.8	8.8	8.8	17.5	7.01	8.77	1.75	1.75	3.51
150	8	2	1	10.2	150	150	20	139.4	2.32-0.31	5.57-9.29	6.00-10.00	16.4	20.4	4.1	4.1	8.2	2.45	3.07	0.61	0.61	1.23
200	22	0	2	21.9	150	150	20	137	2.28-0.30	5.48-9.13	6.00-10.00	35.1	43.8	8.8	8.8	17.5	7.01	8.77	1.75	1.75	3.51
150	9	0	15	14.0	30	20	20	17.52	0.05-0.01	0.22-0.44	0.38-0.75	1.6	2.0	0.4	0.4	0.8	0.24	0.30	0.06	0.06	0.12
365	9	0	15	19.0	30	20	20	23.76	0.07-0.02	0.30-0.59	0.38-0.75	2.1	2.7	0.5	0.5	1.1	0.78	0.98	0.20	0.20	0.39
250	2	3	9	9.6	10	10	5	6.857	0.27-0.11	0.34-0.69	0.50-1.00	1.4	1.8	0.4	0.4	0.7	0.36	0.45	0.09	0.09	0.18
365	4	0	20	7.0	30	10	5	8.758	0.35-0.15	0.44-0.88	1.50-3.00	3.2	3.9	0.8	0.8	1.6	1.15	1.44	0.29	0.29	0.58
250	1	3	10	14.0	5	5	5	5	0.20-0.08	0.25-0.50	0.25-0.50	1.1	1.3	0.3	0.3	0.5	0.3	0.3	0.1	0.1	0.1
	1	3	10	14.0	5	5	5	5	0.20-0.08												
250	1	3	10	14.0	150	150	150	150	6.00-2.50	7.50-15.00	7.50-15.00	31.5	39.4	7.9	7.9	15.8	7.9	9.8	2.0	2.0	3.9
	1	3	10	14.0	150	150	150	150	6.00-2.50												
250	8	3	3	11.1	20	10	10	15.89	0.53-0.10	1.00-2.00	1.00-2.00	3.3	4.2	0.8	0.8	1.7	0.83	1.04	0.21	0.21	0.42
	1	3	3	4.0	20	10	10	11.51	3.84-0.72	5.75-11.51	10.00-20.00	12.1	15.1	3.0	3.0	6.0	3.02	3.78	0.76	0.76	1.51
250	4	2	4	7.0	20	10	10	14	1.17-0.47	2.33-3.50	3.33-5.00	5.8	7.3	1.5	1.5	2.9	1.46	1.82	0.36	0.36	0.73
365	1	0	7	8.0	20	20	20	20	0.67-0.17	0.67-2.00	0.67-2.00	2.1	2.7	0.5	0.5	1.1	0.78	0.97	0.19	0.19	0.39

Tables 4 shows the specific characteristics for mechanical ventilation.

The number of operating days of mechanical ventilation has been reduced by the number of days on which the external weather conditions permit natural ventilation, provided this is suitable for the type of use. The operating hours per day and the fresh-air flow volumes are subdivided into full, partial, and base load. Depending on the respective operating mode, use-specific annual average values can be calculated for the air changes and flow volumes required to create good hygienic air conditions. The figures quoted here are intended to reflect optimum values for hygiene, comfort, and energy efficiency. Standard values and performance regulations tend to err on the higher side and can lead to oversized installations. The listed range of daily peak loads includes values which, in energy performance requirements, are sometimes referred to as normal load, and this therefore leads to a significant increase in energy consumption through ventilation and the heat loss caused by it.

Depending on the design of the ventilation system (just air extraction or air intake, and air extraction, with or without heat recovery, with or without humidity recovery, with minimum or maximum filtering) and the above-listed characteristics of the zone, it is possible to estimate the electricity demand both for the type of use and the type of installation.

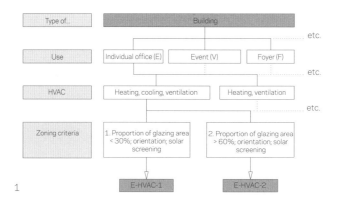

1

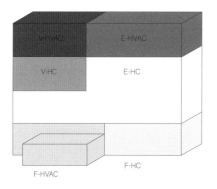

2

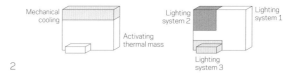

3

1 Structuring a building by the type of use, the type of heating/air conditioning and the design of the building envelope.
In the first step, we divide the building into zones based on their use profiles with their different internal factors. Once a zone has been assigned a use profile, the type and intensity of its use can be described in figures. In the second step, we consider the services installations and hence the type of heating/air conditioning. In the next step, we further subdivide the building according to the features of each room, such as the proportion of window area and the lighting system.

2 Structuring based on areas needing different services, superimposed with use zones. Independently of the division into use zones (*zoning*), the building is subdivided into "services areas" according to the degree of services provided. A services area comprises all those areas in a building which are serviced with the same type of services. This is done separately for the different services, such as domestic hot water (DHW), heating, cooling, ventilation, lighting, etc., and is independent of the zoning process. It is done for each of the different services and is not part of the zoning.

3 It follows that it is quite possible that the boundaries of the zones do not coincide with those of the services areas. In that case, a further subdivision is carried out. However, the boundaries of small services areas often coincide with certain use zones, for example a decentralized cooling system may well coincide with the area of a meeting room. On the other hand, it is also possible that one services area, for example the central heating, covers all use zones in a building if none of the rooms/areas needs a different type of heating or air conditioning system. It is quite possible that the different types of services areas have different boundaries, but no further subdivisions should be carried out for reasons of clarity. The division into zones and the definition of services areas are separate activities that are carried out in parallel.

The objective of this exercise is to achieve greater clarity regarding the different zones and the different types of services but also to identify any potentially critical areas, for example rooms with a large proportion of window area but no active ventilation and cooling.

BUILDING AND TECHNOLOGY

4 Aperture area (effective collector surface without frame) of solar thermal collectors and monthly contribution to annual yield. Depending upon the aperture area per person, solar thermal systems are classed as small installations purely for supplying domestic hot water (DHW), or as larger combination systems for preparing DHW and additional support of the heating system. Since solar radiation varies with the seasons, the yield also varies throughout the year, with the highest output achieved in summer. In winter the aperture area of a small system is not sufficient to generate a significant yield. On the other hand, the larger combination systems are not fully utilized during summer as there is little requirement for space heating. This means that they produce unused surplus. These systems are at their most efficient in spring and autumn when there is a requirement for space heating as well as sufficient solar radiation. For this reason, the tilt of combination systems should suit the lower position of the sun during spring and autumn.

5 Correction factors for tilt and angle of deviation from due south for small solar thermal systems used solely for preparing domestic hot water.
Values are based on a system with a tilt of 45° facing due south. The factors indicate that deviations in orientation and tilt only lead to small reductions in yield compared to the position with the optimum energy yield. There is a clear advantage to architecturally integrating such a system into the roof structure as the reductions in yield are comparatively small.

6 Correction factors for tilt and angle of deviation from due south for combined solar collectors for preparing DHW and supporting the space heating system.
A comparison of the reduction factors for solar thermal systems for DHW preparation only with those giving additional support for the space heating system shows the latter are well suited to being integrated in vertical facades (90°) provided that the deviation from due south is not excessive. With regard to vacuum tube collectors, when the collectors are installed in a vertical position, on a wall or parapet, the tubes themselves can be tilted upwards towards the sky by 15 to 30°.

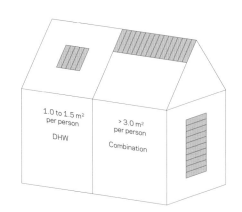

Month	DHW systems	Combi systems
January	2.7%	5.7%
February	3.2%	5.5%
March	6.9%	11.7%
April	13.7%	13.4%
May	12.3%	10.9%
June	13.4%	8.1%
July	14.2%	8.7%
August	11.8%	7.3%
September	10.7%	9.8%
October	7.0%	9.7%
November	3.3%	6.9%
December	0.8%	2.3%

4   DIN V 18599-8:2007-02. Table 10, showing the monthly contribution to the annual DHW yield; DIN V 18599-5:2007-02. Table 18, showing the monthly contribution to the annual yield of a combination system

Tilt factors		Angle of deviation from due south								
		East: γ = -90°				West: γ = +90°				
		-60°	-40°	-20°	0°	20°	40°	60°	90°
Tilt	0°	0.81	0.81	0.81	0.81	0.81	0.81	0.81	0.81
	15°	0.8	0.86	0.88	0.90	0.91	0.91	0.9	0.87	0.81
	30°	0.79	0.88	0.93	0.96	0.98	0.97	0.95	0.91	0.83
	45°	0.76	0.88	0.94	0.98	1.00	1.00	0.97	0.93	0.82
	60°	0.72	0.85	0.91	0.95	0.98	1.00	0.95	0.91	0.79
	75°	0.65	0.78	0.81	0.87	0.89	0.89	0.88	0.85	0.72
	90°	0.54	0.66	0.68	0.69	0.71	0.73	0.75	0.74	0.63

5   DIN V 18599-08:2007-02. Table 11: Correction factors for tilt and orientation (DHW)

Tilt factors		Angle of deviation from due south								
		East: γ = -90°				West: γ = +90°				
		-90°	-60°	-40°	-20°	0°	20°	40°	60°	90°
Tilt	0°	0.66	0.66	0.66	0.66	0.66	0.66	0.66	0.66	0.66
	15°	0.66	0.73	0.76	0.79	0.81	0.80	0.78	0.75	0.68
	30°	0.65	0.78	0.85	0.90	0.93	0.92	0.88	0.82	0.70
	45°	0.64	0.81	0.90	0.97	1.00	0.98	0.94	0.86	0.70
	60°	0.62	0.80	0.91	0.99	1.02	1.01	0.95	0.86	0.68
	75°	0.56	0.76	0.87	0.95	0.99	0.98	0.92	0.82	0.64
	90°	0.49	0.67	0.76	0.83	0.86	0.86	0.82	0.75	0.57

6   DIN V 18599-08:2007-02. Table 19: Correction factors for tilt and orientation (Combi system—heating and DHW)

BIBLIOGRAPHY / PICTURE CREDITS

Achilles, Andreas et al.: Glasklar. Munich 2003

Bauer, Michael; Mösle, Peter; Schwarz, Michael: Green Building: Guidebook for Sustainable Architecture, Munich 2007

Behling, Sophia; Behling, Stefan: Sol Power, Munich 1996

Brand, Silke: Das Meso- und Mikroklima. Contribution to the seminar about climate change, climate history of the planet, and climatology at the University of Tübingen, Tübingen 2008

Daniels, Klaus: Low-Tech Light-Tech High-Tech, Basel 1998

Daniels, Klaus: The Technology of Ecological Building, Basel 1999 (2nd ed.)

Daniels, Klaus; Hammann, Ralph: Energy Design for Tomorrow, Fellbach 2009

DIN (Ed.): DIN V 18599, Part 1 to 10, 2007

DIN (Ed.): DIN V 18599 Teil 100, 2010

EW Medien und Kongresse GmbH (Ed.): RWE Bau-Handbuch, Frankfurt am Main 2010 (14th ed.)

Fachgebiet Energieeffizientes Bauen, EW Medien und Kongresse (Ed.): Vorstellung der Energieeinsparverordnung 2009 content of the correspondence course to become an energy consultant at the TU Darmstadt 2009

Glücklich, Detlef: Ökologisches Bauen, Munich 2005

Häckel, Hans: Meteorologie, Stuttgart 1993

Hausladen, Gerhard et al.: ClimateDesign, Basel 2005

Hausladen, Gerhard et al.: ClimateSkin, Basel 2008

Hegger, Manfred et al.: Energy Manual: Sustainable Architecture, Basel 2008

Hennings, Dr. Detlef (et al.): Leitfaden Elektrische Energie (LE), Darmstadt 2001

Hochbauamt Frankfurt am Main (Ed.), Abt. Energiemanagement: Leitlinien wirtschaftliches Bauen 2010, Frankfurt am Main 2010

Hochberg, Anette; Hafke, Jan-Henrich; Raab, Joachim: Open I Close, Basel 2009

Hupfer, Peter (Ed.): Das Klimasystem der Erde, Berlin 1991

Katzschner, Lutz: Designing Open Spaces in the Urban Environment, 2004; in: Nikoplolou, M. CRES, Fiths framework programe EU; RUROS, p. 22–26

Lenz, Bernhard; Schreiber, Jürgen; Stark, Thomas: Sustainable Building Services, Munich 2010

Pistohl, Wolfram: Handbuch der Gebäudetechnik, München 2005

Roggel, Klaus: Bauphysikalische Erläuterungen, Berlin 2000

Voss, Karsten et al.: Bürogebäude mit Zukunft, Berlin 2006 (2nd ed.)

Weischet, Wolfgang; Endlicher, Wilfried: Einführung in die Allgemeine Klimatologie, Stuttgart 2008

Willems, Wolfgang: Sommerlicher Wärmeschutz. Lecture series at the Ruhr University Bochum 2008

Photos without credits are from the authors/editors. Despite intense efforts, it was not possible to identify or contact the copyright owners of certain photos. Their rights remain unaffected, however, and we request them to contact us.

Baan, Iwan: p. 13 2
Bauatelier Schmelz & Partner: p. 91 3
Behnisch, Stefan: p. 113 2
Blaser, Chrsitine: p. 85
Bühler, Beat: p. 75 5,6
Crocker, Tim: p. 127 1
Esch, Hans Georg: p. 13 2
Furer, René: p. 85
Halbe, Roland: p. 91 4
Hart, Rob't: p. 85; p. 132–135
Hartwig, Joost: p. 29 7a; p. 87 3, 4
Hergesell, Tanja: p. 85
HG Esch for Ecophon: p. 11 2
Hopp, Florian: p. 24 1g
Jarmund/Vigsnæs AS Arkitekter MNAL (photo: Nils Petter Dale): p. 22
kämpfen für architektur ag: p. 136; p. 137 3
Kane, Nick: p. 124–126; p. 127 2
Katzschner, Lutz: p. 31 6,7
Keller, Michael: p. 91 5; p. 102 2; p. 113 3

Korte, Niko: p. 85
Kracher, Willi: p. 137 1 2; p. 138; p. 139
Mahal, Nikola/Bischoff, Björn: p. 18–21
Mørk, Adam: p. 86 2
Olson Kundig Architects (photo: Tim Bies): p. 40
Ott, Thomas: p. 88; p. 91 6
Ouwerker, Erik-Jan: p. 27
Perret, Christian: p. 10 1
Rainer, Simon: p. 64; p. 128–131
Robson, Jonathan: p. 14
Schiess, Hanspeter: p. 122; p. 140–143
Schmidt, Leon: p. 93 2
Smith, Melissa K.: p. 66; p. 85
Tetro, Jim: p. 93 3
Traxler, Stefan: p. 42 1
Ventomaxx GmbH: p. 102 1
Willebrand, Jens: p. 12 1
Zumthor, Peter (photo: Helene Binet): p. 49

PUBLISHING INFORMATION

Editors: Alexander Reichel, Kerstin Schultz
Concept: Alexander Reichel, Kerstin Schultz, Andrea Wiegelmann
Authors: Manfred Hegger, Joost Hartwig, Michael Keller
Authors' Assistants: Katharina Baumann, Erhan Ekizoglu, Katharina Fey,
Isabelle von Keitz, Kristina Klenner, Hui Hui Lü, Nikola Mahal

Editorial Supervision: Andrea Wiegelmann, Katharina Sommer

Translation from German into English: Hartwin Busch
Copy Editing: Raymond Peat
Proofreading: David Koralek

Drawings: Julia Nellessen, Anna Tomm
Layout and Cover Design: Nadine Rinderer
Typesetting: ActarBirkhäuserPro

The technical and construction recommendations contained in this book are based on
the present state of technical knowledge. They should be checked in each case against
the relevant instructions, standards, laws etc. as well as local regulations before applying
them. No liability is accepted.

A CIP catalogue record for this book is available from the Library of Congress, Washington DC, USA.

Bibliographic information published by the German National Library.
The German National Library lists this publication in the Deutsche Nationalbibliografie;
detailed bibliographic data are available on the Internet at http://dnb.d-nb.de.

This work is subject to copyright. All rights are reserved, whether the whole or part of the
material is concerned, specifically the rights of translation, reprinting, re-use of illustrations, recitation, broadcasting, reproduction on microfilms or in other ways, and storage
in data bases. For any kind of use, permission of the copyright owner must be obtained.

This book is also available in a German language edition (ISBN 978-3-0346-0511-3).

© 2012 Birkhäuser, Basel
P.O. Box, CH-4002 Basel, Switzerland
Part of De Gruyter

Printed on acid-free paper produced from chlorine-free pulp. TCF ∞

Printed in Germany

ISBN 978-3-0346-0513-7

9 8 7 6 5 4 3 2 1   www.birkhauser.com